Historical Biogeography of the Southeast Asian genus Spatholobus (Legum.–Papilionoideae) and its allies

PROEFSCHRIFT

TER VERKRIJGING VAN DE GRAAD VAN DOCTOR
AAN DE RIJKSUNIVERSITEIT TE LEIDEN,
OP GEZAG VAN DE RECTOR MAGNIFICUS DR. L. LEERTOUWER,
HOOGLERAAR IN DE FACULTEIT DER GODGELEERDHEID,
VOLGENS BESLUIT VAN HET COLLEGE VAN DEKANEN
TE VERDEDIGEN OP MAANDAG 4 NOVEMBER 1996
TE KLOKKE 14.15 UUR

DOOR

JEANNETTE WILMA ABÉLE RIDDER-NUMAN
geboren te Emmen in 1957

PROMOTIECOMMISSIE:

Promotor: Prof. dr. C. Kalkman
Co-promotor: Dr. M.C. Roos

Referent: Dr. R.M. Polhill (Kew, Engeland)

Overige leden: Prof. dr. P. Baas
                       Prof. dr. E. Gittenberger
                       Prof. dr. ir. L.J.G. van der Maesen (LU Wageningen)

Rijksherbarium / Hortus Botanicus, Leiden University, The Netherlands.

The research for this thesis was supported by grant No. 40.041 from the 'Nederlandse Organisatie voor Wetenschappelijk Onderzoek / Stichting Levenswetenschappen'.

This thesis has (except for Chapter 5) also been published as:
J.W.A. Ridder-Numan. 1996. The historical biogeography of the Southeast Asian genus *Spatholobus* (Legum.–Papilionoideae) and its allies. Blumea Supplement 10: 1–144, illus. ISSN 0006-5196, ISBN 90-71236-30-7.

Opgedragen aan
Jan, voor zijn steun en geduld
Pieter, Mark en Bouke
en mijn ouders

CONTENTS

SUMMARY

The genera *Butea*, *Meizotropis* and *Spatholobus* (Leguminosae–Papilionoideae), occurring in continental Southeast Asia and the West Malesian Archipelago, are treated phylogenetically with *Kunstleria* as an outgroup. The genera *Spatholobus* (29 species), *Butea* and *Meizotropis* (each 2 species) are usually placed in the tribe Phaseoleae of the Papilionoideae, whereas *Kunstleria* belongs to the less advanced tribe Millettieae. *Spatholobus* is probably basal in the Phaseoleae. An analysis with PAUP of the datamatrix containing 80 macromorphological, 10 leaf anatomical and 7 pollen morphological characters resulted in three most parsimonious trees (MPTs) with a length of 589 steps. As for the method, it turned out that the option 'addition sequence random' gave more MPTs than an 'addition sequence simple'. The standard option 'addition sequence simple' with taxon one as reference taxon resulted in trees a few steps longer than the MPTs.

One of the three phylogenetic trees has been chosen to be used for the biogeographic analysis. Other genera used for this analysis are *Fordia* (Leguminosae–Papilionoideae), *Genianthus* (Asclepiadaceae), and *Xanthophytum* (Rubiaceae). Within Southeast Asia 29 areas of distribution were recognised. Although they are solely based on the distribution pattern of the species, some areas coincide with geological entities, e. g., E Malaya and SE Sumatra. The biogeographical analysis performed by PAUP (BPA) and CAFCA (CCA) resulted for PAUP in 19 MPTs under assumption 0 with a length of 366 steps and 324 MPTs under assumption 1 with a length of 316 steps; for CAFCA the analysis resulted in 1 MPT (450 steps) under assumption 0 and 1 MPT (417 steps) under assumption 1. In all results more or less the same larger groups of areas were present: the areas recognised on Borneo, those on the Malay Peninsula together with SE Sumatra and W Java, and the areas on the continent of Southeast Asia. Some areas with only one species were found basal to the other nodes, and were considered not informative. After comparing the 19 MPTs, the 50% Majority-rule consensus tree was used to discuss the possible link with the geology. The geology of the region is very complex as there are four major tectonic plates in collision with each other: the Eurasian Plate, the Indo-Australian Plate, the Pacific Plate, and the Philippines Sea Plate.

A summary of the geology of Southeast Asia is given. It is probable that the general area-cladogram reflects the split (during high sea level) between continental Southeast Asia and Peninsular Malaysia and the Malay Archipelago. Furthermore it is remarkable that the areas in Borneo are splitting off in the order of age, where the youngest areas are at the top and the oldest at the base of the tree. The history of the genus *Spatholobus* is hypothesised by backtracking the phylogenetic relationships on the general area-cladogram. Probably the history of *Spatholobus* was influenced by the differences in sea level. During low sea levels the Sundaland Plateau was dry and it was possible to migrate into the Malesian Archipelago, but high sea levels resulted in isolation and speciation. It is impossible to say more about timing than that it could have started at the earliest in the Early Eocene. In the last case the ancestor of the genus was present in an area without the northern parts of Borneo, and with the south arm of Sulawesi still connected to Borneo.

Chapter 1

INTRODUCTION

The aim of this study is to reconstruct the phylogeny of the genus *Spatholobus* (and allies) and to use this phylogeny in a historical biogeographical analysis. This project is part of an ongoing research programme on the historical biogeography of the Malesian region at the Rijksherbarium / Hortus Botanicus Leiden. Other groups already studied are the sapindaceous genera *Cupaniopsis* (Adema, 1991), *Guioa* (Van Welzen, 1989), *Arytera* and *Mischarytera* (Turner, 1995), and sections from the orchid genus *Bulbophyllum* (Vermeulen, 1993). These studies focus on the East Malesian region, whereas this study will go into more detail on the West Malesian region including the Southeast Asian continent.

I will not add to the discussion on the methods and ideas behind phylogenetics, as there is a wealth of information on methodology and already many practical studies have been published (Brooks, 1990; Brooks & McLennan, 1991; Cracraft, 1983, 1988; Forey et al., 1993; Nelson & Platnick, 1981; Page, 1988, 1990; Wiley, 1981, 1987; Zandee & Roos, 1987). Several computer programs are available for use with morphological or molecular data, e.g., PAUP (Swofford, 1991, 1993) Hennig 86 (Farris, 1988), and CAFCA (Zandee, 1995).

For a phylogenetic treatment it is necessary to assume the monophyly of the group under study. In addition it is necessary to include an outgroup, to polarise the character states used in the phylogenetic analysis. The computer program CAFCA can run without the use of an outgroup.

The legume genus *Spatholobus* consists of 29 species of lianas in Southeast Asia (Ridder-Numan & Wiriadinata, 1985; Ridder-Numan, 1992). The closely allied genera *Butea* and *Meizotropis* occur only on the mainland of Southeast Asia. Both comprise two species (Sanjappa, 1987). The two species of *Butea* have large red or yellow flowers, and are probably bird-pollinated. One species, *B. monosperma*, is a tree, the other is a liana. The genus *Meizotropis* occurs in the Himalayan region and is a shrub. All three genera are usually placed in the tribe Phaseoleae (Lackey, 1981). On account of the resemblance in mainly pod and leaflets, however, these three genera have also been treated as one genus, *Butea* [e.g., Blatter (1929), and in several regional Floras]. The pod is one-seeded and samaroid; the leaves are trifoliolate and asymmetric, as in most Phaseoleae (Lackey, 1981). Not all species of *Spatholobus* closely resemble *Butea*, since in *Spatholobus* there are, e.g., two types of leaflets: strongly asymmetrical leaflets (Fig. 2.8 and 2.10), as is usual in the tribe Phaseoleae, and more or less symmetrical leaflets (Fig. 2.2), as is usual in the Millettieae. This last tribe is considered to be less derived than the Phaseoleae (Geesink, 1981, 1984). A basal position in the tribe Phaseoleae is indicated also by a phylogenetic analysis that is based on molecular data (Bruneau et al., 1995). Here *Spatholobus* has a position low in the tribe Phaseoleae as sister taxon to *Butea*.

Kunstleria (Fig. 2.2 and 2.20b) has been chosen as an outgroup. This genus with eight species in the tribe Millettieae bears a great resemblance to *Spatholobus*. The flowers are of the same size and appearance, and have the same number of ovules (two). The leaflets of *Kunstleria* are more or less symmetrical. Some species have uni- or trifoliolate leaves, others have more than three leaflets. *Kunstleria* differs from *Spatholobus* mainly in the vexillary stamen, which is adnate to the claw of the standard in *Kunstleria* (versus free in *Spatholobus*), the extraordinary long and often curled dorsal auricle on the wing petals, and the flat strap-like pod with the seed(s) placed centrally (Ridder-Numan & Kornet, 1994; Fig. 2.20b). Unfortunately no molecular data on this rare liana genus are available. *Kunstleria* has been mentioned in molecular treatments, but this appeared to be *K. blackii*, which is synonymous with *Austrosteenisia blackii*. *Austrosteenisia* is a genus probably related to the less derived genera in the Millettieae, i.e. *Callerya* and *Ostryocarpus*, and it occurs in Australia and New Guinea (Geesink, 1984).

The phylogenetic treatment of *Spatholobus* is given in **Chapter 2**. The characters used for the phylogenetic analysis are discussed in this chapter in the section on morphology. They include macromorphological characters as well as some from leaf anatomy and pollen morphology. The pollen morphological characters were selected after an extensive review of the pollen morphology of *Spatholobus*, *Butea*, *Meizotropis* and *Kunstleria*, which will be published separately (Ridder-Numan, in press; in the thesis added as **Chapter 5**). The result of the phylogenetic analysis has been used in the biogeographical analyses.

The historical biogeographical analysis serves two purposes: to construct a generalised area cladogram providing information on the biogeographic history of Southeast Asia, and to explain the developments of the genus *Spatholobus* in space and time in the area as a special case. Of course without any knowledge of the geology of the region it is impossible to interpret a generalised area cladogram other than as a sequence of areas. Combining this area sequence with the geological background may give an indication of geological events causing cladogenesis. It is difficult to establish the order of these events unless there are either well-dated events recognisable in the area cladogram, or there is a complete fossil record of the group. As for fossils, there are none available as far as I know from the group under study. It can only be assumed that the vicariance events which caused speciation in this group occurred after the existence of the (sub)family in the area. **Chapter 3** gives an overview of the geological knowledge on the Southeast Asian region. The general information available on fossils of Leguminosae in the region is presented.

Chapter 4 deals with the historical biogeographical analyses. The basis of these analyses are the phylogenies of the genera *Spatholobus* and allies (*Chapter 2*), *Fordia* (Schot, 1991), *Genianthus* (Klackenberg, 1995), and *Xanthophytum* (Axelius, 1990). Phylogenies of these genera, which are distributed in more or less the same region as *Spatholobus*, were available at the time of this study. With their combined phylogenies and distributions a generalised area cladogram has been established, after analysis with CAFCA and PAUP, computer programs appropriate for use in cladistic biogeographic analysis. The result of these analyses, the generalised area cladogram, is compared with the geological information. The genus *Spatholobus* is optimised on the area cladogram, and its possible history is given.

Acknowledgements

Very helpful in reading of and commenting on the various stages of (parts of) the manuscript were: Dr. A. Davis, Dr. I. K. Ferguson, Dr. R. M. Polhill, Dr. B. D. Schrire, Prof. Dr. H. Wensink, Dr. W. van der Werff.

Chapter 2

PHYLOGENETIC ANALYSIS

INTRODUCTION

The legume genus *Spatholobus* (Ridder-Numan & Wiriadinata, 1985; Ridder-Numan, 1992), and the closely allied genera *Butea* and *Meizotropis* (Sanjappa, 1987) are used as an ingroup for the phylogenetic analysis. *Spatholobus* and its allies are basal in the tribe Phaseoleae (Bruneau et al., 1995). The genus *Kunstleria*, which resembles *Spatholobus*, is chosen as an outgroup. This latter genus belongs to the tribe Millettieae (Geesink, 1981, 1984). The Millettieae are a complex paraphyletic group, and the last to be revised for Flora Malesiana. The genera are chosen to study the historical bio-geography of the West Malesian area, and – in addition – the choice of these genera may add to a better understanding of the Millettieae.

MATERIAL AND METHODS

Introduction

A data matrix with 97 characters, 7 of which relate to pollen morphology, 10 to leaf anatomy, and 80 to macromorphological characters, is presented for all species of *Spatholobus*, *Butea* and *Meizotropis* (Table 2.1). The characters and their states are described in the section on Morphology, and the quantitative characters are given in Table 2.2. The outgroup *Kunstleria* was considered a single taxon by coding the combined information on the species as is otherwise done for specimens in describing a species. Autapomorphic character states occurring in only one species were ignored, and the state occurring in the rest of the genus was taken as representative for the genus.

Data for the matrix were taken from herbarium specimens. Where possible, samples of pollen or leaves were taken, preferably from 10 samples per species. A list of specimens used for the pollen morphological samples can be found in Ridder-Numan (in press). Specimens used for leaf anatomical sections are listed in the Appendix to the present chapter.

Method of analysis

The data matrix was analysed using the computer program PAUP (Swofford, 1993) under the heuristic search option with a stepwise addition sequence either simple or random, and the branch-swapping method tree-bisection-reconnection (TBR). All characters were analysed unordered; uninformative characters, e. g., autapomorphies, were ignored and characters with multiple states were treated as polymorphisms. The possibility of using multistate characters that are treated as polymorphisms was one of the reasons for using PAUP. It must be noted, however, that PAUP counts steps differently from other programs, especially when polymorphisms are involved. Polymorphic characters contain enough phylogenetic information to be included in a phylogenetic analysis (Wiens, 1995). In one stage or another characters must have been polymorphic during species development. To avoid too much homoplasy due to the increased variability by polymorphy, it is perhaps best to keep the number of states for one character as low as possible. I did this by re-evaluating the character states using the program MacClade (Maddison & Maddison, 1992), and recoding the characters where possible. Polymorphic species were coded as such, i.e. taxa coded as having multiple states for a particular character were not treated as uncertain (= '?') for that character, as is often necessary in other computer programs. In theory polymorphic species should be coded following the locally plesiomorphic state. This state, however, is not known in advance, and should be derived from outgroup comparison if possible. For a detailed discussion on coding of polymorphic species see Turner (1995).

A search with a random addition sequence was performed to find more solutions than with the simple addition sequence. As described in more detail below, it became evident that with a simple addition sequence not all islands of trees were found. This may be caused by the different trees created initially when different taxa are chosen as reference taxa (a simple addition sequence needs a reference taxon). After tree bisection and reconnection (TBR) these different initial trees may lead eventually to different islands of trees (Swofford, 1991). By creating more starting trees, a random addition sequence is more successful in finding more islands of trees (and in that way more parsimonious trees).

The default option for running a TBR search is a simple addition sequence and has as reference taxon 1, which will be in most cases the outgroup. By assigning each time a different member of the taxa in the analysis as reference taxon and performing a new TBR search different trees appeared, which were in some cases shorter. Using different taxa as reference taxon in a TBR search thus led to different results. To explore more possible solutions (by creating more different starting points and thus possibly different islands of trees) an analysis with addition sequence random (1000 repetitions) was performed. This analysis resulted in one additional most parsimonious tree. The first two most parsimonious trees (MPT) were found during the first replicate, the additional one during the fifteenth. During the other 85 replications no additional trees were found.

Finally the most parsimonious trees that resulted from the search with PAUP were analysed using MacClade (Maddison & Maddison, 1992) with several options such as Trace Character, Trace All Changes, Compare Two Trees.

MORPHOLOGY

The characters used for the phylogenetic analysis were taken from various parts of the plant. Although revisions of the genera were available (Ridder-Numan & Wiriadinata, 1985; Sanjappa, 1987; Ridder-Numan, 1992; Ridder-Numan & Kornet, 1994), all characters were checked against material available. Due to the fact that more material was available than at the time of the revision and that the revisions were not written with the intention of a phylogenetic treatment, it turned out that not all data presented in these revisions were detailed enough, and, moreover, that not all characters I wanted to use for the analysis could be found in the descriptions. In addition to macromorphological characters I checked pollen morphology as well as leaf anatomy for characters useful for a phylogenetic analysis. Due to the limitations of the herbarium material the characters selected are restricted to those parts which are generally collected from lianas (or trees), i.e. the terminal parts of the plants where the inflorescences may be found. Features of the older stems and leaves are not discussed as they are not available for most of the species. For some species material is very scarce, and consequently the data for these species is based on a small number of specimens only. A description of the characters and states used for the phylogenetic analysis is presented below. An overview of the characters and the states recognised is given in Table 2.1. The numbers between square brackets in the descriptions below refer to the numbers of the characters in this table. For quantitative characters Table 2.2 gives more details. In case of absence of a plant part, a question mark is used for the characters that relate to this part and cannot be scored. Thus, all flower characters are designated as a '?' for species of which no flowers were available. The lack of material was considerable for some species, making it impossible to avoid question marks. Poorly known species can be best omitted. One species, *Spatholobus bracteolatus*, was excluded from the phylogenetic analyses because the data that could be collected were too incomplete.

Table 2.1. Characters and character states used in the phylogenetic analysis.

a. List of characters and character states

Stem

1. Stem hollow
    1 = yes
    2 = no
2. Indumentum
    1 = sparse / not pubescent
    2 = pubescent / puberulous / sericeous
    3 = hirsute / strigose / pilose
3. Lenticels
    1 = not conspicuous / small
    2 = wart-like
    3 = elongated
4. Exudate
    1 = present
    2 = absent
5. Bark
    1 = smooth
    2 = with ridges / wrinkles

Leaves

6. Ultrajugal part of the rachis
    1 = less than 1/10 of the rachis / no ultrajugal part
    2 = ultrajugal part more than 1/10
7. Stipules
    1 = recurving
    2 = not recurving

8. Stipules
    1 = (early) caducous
    2 = more or less persistent
9. Stipules symmetrical
    1 = yes
    2 = no (asymmetrical)
10. Stipules nerves
    1 = nerves visible
    2 = no visible nerves
11. Stipules indumentum
    1 = glabrous / glabrescent
    2 = hairy
12. Stipule length
    1 = as long as wide
    2 = 2 or 3 times as long as wide
    3 = 4 or more times as long as wide
13. Pulvinus colour
    1 = black / darker than rachis
    2 = same colour as rachis
14. Pulvinus indumentum
    1 = glabrous / glabrescent / sparsely hairy
    2 = pubescent / hirsute
    3 = densely pubescent/ strigose
15. Stipellae caducous
    1 = (early) caducous
    2 = more or less persistent

(Leaves continued)

16. Stipellae length
    1 = up to the length of the petiolule
    2 = conspicuously longer than the petiolule
17. Upper surface colour
    1 = green / brownish (dried colour)
    2 = greyish blue (glaucous)
18. Lower surface colour
    1 = green / brownish (dried colour)
    2 = always brown
19. Upper surface indumentum
    1 = glabrous to sparsely pubescent
    2 = pubescent (dense)
    3 = sericeous
    4 = hirsute / strigose
20. Lower surface indumentum
    1 = glabrous to sparsely pubscent
    2 = pubescent (dense)
    3 = sericeous
    4 = hirsute / strigose
21. Shape of top leaflet
    1 = only elliptic
    2 = ovate to elliptic
    3 = obovate to elliptic
22. Apex leaflet
    1 = acute
    2 = acuminate / cuspidate
    3 = emarginate / rounded
    4 = abruptly acuminate
23. Base leaflet
    1 = acute / obtuse / subcordate
    2 = decurrent
24. Lateral leaflet symmetrical
    1 = more or less symmetrical
    2 = strongly asymmetrical
25. Main nerve
    1 = sunken or flat
    2 = raised
26. Secondary nerves at upper surface
    1 = flat
    2 = raised
27. Nerves at lower surface
    1 = only midrib prominent
    2 = midrib and secondary nerves prominent
    3 = venation also prominent
28. Venation
    1 = reticulate
    2 = reticulate-scalariform
    3 = scalariform
29. Angle secondary nerves with midrib (average)
    1 = up to 45°
    2 = more than 45°
30. Secondary nerves terminate
    1 = diffusely
    2 = in the margin
    3 = forming marginal arches
31. Pairs of secondary nerves (average)
    1 = 5–8
    2 = more than 8

Inflorescence

32. Place of inflorescence
    1 = axillary / terminally
    2 = terminally

(Inflorescence continued)

33. Indumentum of inflorescence
    1 = puberulous / pubescent
    2 = hirsute / strigose / tomentose
    3 = (nearly) glabrous
34. Number of branches
    1 = main axis with (secondary) side branches / fascicles
    2 = main axis with secondary and tertiary branches / fascicles
    3 = main axis with secondary, tertiary and quaternary branches / fascicles
35. Brachyblasts
    1 = present
    2 = absent, flowers solitary
36. Number of bracts of the first order (= secondary) side branch
    1 = one bract
    2 = two bracts
37. Shape of the bracts
    1 = linear
    2 = triangular
    3 = broadly ovate
    4 = elliptic, surrounding the bud
38. Number of bracts of the tertiary branch / flower
    1 = one bract
    2 = two bracts

Flower

39. Length of the pedicel
    1 = very short, up to 1 mm; flower nearly sessile
    2 = pedicel > 1 mm; flower not nearly sessile
40. Place of the bracteoles
    1 = immediately below the calyx
    2 = on the upper half of the pedicel
    3 = on the lower half of the pedicel
41. Indumentum of the calyx
    1 = sparsely pubescent / nearly glabrous
    2 = pubescent / sericeous
    3 = hirsute
42. Top of the vexillary lobe
    1 = emarginate
    2 = not emarginate
43. Length of the vexillary lobe
    1 = as long as the cup
    2 = about half the length of the cup
    3 = less than half the length of the cup
44. Shape of the other calyx lobes
    1 = triangular
    2 = truncate / somewhat rounded
    3 = more or less rhomboïd
    4 = with a rounded apex
45. Indumentum on the standard
    1 = present
    2 = absent
46. Apex of the standard
    1 = emarginate
    2 = not emarginate
47. Standard with auricles
    1 = present
    2 = absent
48. Base of the standard blade
    1 = decurrent into the claw
    2 = truncate

(Flower continued)

49. Dorsal auricle on the wing
     1 = present
     2 = absent
50. Ventral auricle on the wing
     1 = present
     2 = absent
51. Wing with lateral pocket
     1 = present
     2 = absent
52. Wing with indumentum outside
     1 = present
     2 = absent
53. Wing with indumentum inside
     1 = present
     2 = absent
54. Wing with indumentum on dorsal auricle / margin
     1 = present
     2 = absent
55. Wing with indumentum on ventral auricle / margin
     1 = present
     2 = absent
56. Keel with a dorsal auricle
     1 = present
     2 = absent
57. Keel with a ventral auricle
     1 = present
     2 = absent
58. Keel with a lateral pocket
     1 = present
     2 = absent
59. Keel with indumentum outside
     1 = present
     2 = absent
60. Keel with indumentum inside
     1 = present
     2 = absent
61. Keel with indumentum on dorsal auricle / margin
     1 = present
     2 = absent
62. Keel with indumentum on ventral auricle / margin
     1 = present
     2 = absent
63. Keel petals
     1 = connate
     2 = free
     3 = overlap
64. Vexillary filament
     1 = free
     2 = connate to the claw of the standard
65. Filaments connate for
     1 = less than half of their length
     2 = more than 3/4 of their length
     3 = more than half, but less than 3/4 of their length
66. Indumentum on filament sheath
     1 = absent
     2 = present
67. Size of the anthers
     1 = less than 0.5 mm
     2 = more than 0.5, but less than 1 mm
     3 = more than 1, but less than 2 mm
     4 = more than 2 mm

(Flower continued)

68. Anthers
     1 = fertile and equal
     2 = all fertile and alternately larger and smaller
     3 = alternately fertile and sterile
69. Nectary glands
     1 = 10
     2 = 5, all equal
     3 = connate, no separate lobes
     4 = 10 thickened filament bases
70. Nectary with indumentum
     1 = absent
     2 = present
71. Indumentum of the ovary
     1 = pubescent / sericeous
     2 = very densely pubescent / woolly
     3 = sparsely hairy
72. Indumentum on the style
     1 = up to halfway
     2 = at the base
     3 = up to at least 3/4
73. Number of ovules
     1 = 2
     2 = 2–4
     3 = more than 4
74. Place of the ovules
     1 = basal
     2 = in the middle / apical
75. Stipe of the ovary present
     1 = no
     2 = yes
76. Stigma
     1 = flat
     2 = capitate
     3 = elongated

Pod

77. Stipe of pod exceeding calyx
     1 = yes
     2 = no
78. Top of the pod
     1 = with the style remnant pointing to the dorsal side
     2 = with the style remnant pointing forward
     3 = with the style remnant bent downwards
79. Shape of the wing
     1 = straight, as narrow as the seed-bearing part
     2 = wider than the seed-bearing part
80. Indumentum of pod
     1 = puberulous or nearly glabrous
     2 = densely pubescent

Leaf anatomy

81. Hair base
     1 = basal cell not in palisade layer
     2 = basal cell in palisade layer
82. Bulbous septate hair base
     1 = absent
     2 = present
83. Glandular hair
     1 = stalk 1–2, head 2–6 cells
     2 = stalk 2–5, head 4–8 cells
     3 = head cup-shaped

(Leaf anatomy continued)

84. Epidermis lower surface
    1 = straight
    2 = curved cell walls
    3 = undulate
    4 = strongly undulate

85. Hypodermis
    1 = absent
    2 = only near larger veins
    3 = present throughout leaves

86. Stomata adaxial
    1 = absent
    2 = present

87. Vascular bundle in midrib
    1 = open
    2 = closed / semi-closed

88. Veins transcurrent (larger veins)
    1 = absent (smaller palisade layer continuous)
    2 = present

89. Tannin (?) containing cell layer between phloem and xylem
    1 = present
    2 = absent
    3 = in xylem itself

90. Tannin in bundle sheath
    1 = present
    2 = absent

Pollen

91. Ornamentation pollen
    1 = closed ornamentation (rugulate, psilate-perforate, fossulate)
    2 = microreticulate / coarse perforate
    3 = verrucate

92. Fused colpi
    1 = absent
    2 = yes, sometimes
    3 = often

93. Foot layer / endexine ratio
    1 = more than 0.6
    2 = less than 0.6 and more than 0.1
    3 = less than 0.1

94. Colpus ends
    1 = acute
    2 = obtuse

95. Mesocolpial pouches
    1 = present
    2 = absent

96. Infratectal layer
    1 = well-spaced columellae
    2 = irregular (short) columellae
    3 = granules
    4 = indistinct / dense

97. A/E
    1 = mean < 0.28, max 0.31
    2 = mean 0.28, but < 0.4
    3 = mean > 0.4

b. Data matrix

	1	2	3	4	5	6	7	8	9	10	11	12	13	14	15	16	17	18	19	20
S. acuminatus	2	1/2	2	1	2	2	2	1	1	3	2	3	2	2	1	1	1	1	1	1
S. albus	2	1	2	1	1	1/2	1	1	1	3	2	1	2	1	1	1	2	2	1	1
S. apoensis	2	1	1	2	1	2	2	1	1	3	1	2	1	1	1	1	2	2	1	1
S. auricomus	2	2	1	2	1	2	2	1	1	?	?	?	2	3	1	2	1	2	1	3
S. auritus	1	2	1	1	1/2	2	2	2	1	1	2	2	2	2	1	2	1	1	1	2
S. bracteolatus	1	3	1	2	1	2	2	1	1	3	2	2	1/2	3	1	1	1	1	1	1
S. crassifolius	2	1	1	1	1	2	?	1	?	?	?	?	2	1	1	1	1	1	1	1
S. dubius	2	1	3	1	1	2	2	1	1	1	2	2	2	2	2	1	2	2	1	2
S. ferrugineus	1/2	2	1	1	1	2	1	1	1	3	2	1	2	3	1	1	1	1	1	2/3/4
S. gyrocarpus	2	2	1	1	1	2	?	1	?	?	?	?	2	3	1	1	1	1	1	2
S. harmandii	2	1	1	2	1	2	2	1	1	2	2	2	3	2	1	2	2	1	1	1
S. hirsutus	1	1	1	2	1	2	2	2	1	1	1	3	1	2	1	1/2	1	1	1	1
S. latibractea	2	2	3	2	1	2	2	2	1	1	2	2	2	2	2	1	2	2	1	2
S. latistipulus	2	1	1	1	1	2	2	2	2	1	1	1	1	1	2	1	2	2	1	1
S. littoralis	2	1	3	1/2	1/2	2	2	1	1	3	1	2	2	1	1	1	1	1	1	1
S. macropterus	1/2	1	1	2	2	2	2	1	1	1	1	3	1	1	1	1	1	1	1	1
S. maingayi	2	1	1	2	1	2	2	1	1	1	1	2	1	1	1	1	1	1	1	1
S. merguensis	2	1	1	2	1	1	2	1	1	1	2	2	2	1	1	1	1/2	1	1	1
S. multiflorus	2	2	1	1	1	1	2	2	1	3	2	2	2	3	2	2	2	2	1	3
S. oblongifolius	1	1	1	2	2	1	?	1	?	?	?	?	1	1	1	1	1	1	1	1
S. parviflorus	2	2	1	1	2	2	1	1	1	3	2	1	2	3	1	1	1	1	1	1/2/3
S. persicinus	2	2	1	1	1	2	2	1	1	3	2	2	2	2	1	1	1	2	1	2
S. pottingeri	1	3	1	1/2	2	2	1/2	1	1	3	2	2	1/2	1	2	1/2	1	1	1	1/4
S. pulcher	2	1/3	3	2	1	2	?	1	?	?	?	?	2	2	1	1	1	1	1	4
S. purpureus	2	1	1	2	2	2	?	1	?	?	?	?	2	2	1	1	1/2	1	1	1
S. ridleyi	2	1	1	2	2	2	2	1	1	1	1	2	1	1	2	1	1	1	1	1
S. sanguineus	2	2	1	1	1	2	1	1	1	3	2	2	2	3	1	2	1	1	1	2
S. suberectus	2	1	2	1	2	2	2	1	1	3	2	2	2	1	1	1	1	1	1	1
S. viridis	2	2	2	2	1	2	2	1	1	1	2	2	2	2	1	1	1	1	1	2
B. monosperma	1	2	1	1	2	2	1	1	1	3	2	2	2	2	1	1	1	1	1/3	3
B. superba	1	2	1	1	2	2	1	1	1	3	2	2	2	2	1	1	1	1	2/3	2
M. buteiformis	2	2	1	1	2	2	2	1	1	1	2	1	2	2	1	1	1	1	2	2
M. pellita	1	3	1	1	2	2	2	1	1	3	2	1	2	3	1	1	1	1	4	4
Kunstleria	1	2	2	2	2	2	2	1	1	3	2	2	2	2	?	?	1	1	1	2

(Table 2.1 contd)	21	22	23	24	25	26	27	28	29	30	31	32	33	34	35	36	37	38	39	40
S. acuminatus	1	1	1	1	1	1	1	1	2	1	1	1	1	1	2	2	2	1	2	3
S. albus	1	1	1	1	1	1	1	1	1	1	1	1	1	1	2	1	2	1	2	1
S. apoensis	1	4	1	1	1	1	2	1	2	1	1	1	1	2	2	2	2	1	2	2
S. auricomus	1	2	1	1	1/2	1	2	1	1	1	1	1	1	2	1/2	1	3	1	2	1
S. auritus	1	2	1	1	1	1	3	2	1	3	2	1	1	2	1	2	1	1	2	1
S. bracteolatus	1	2	1	2	2	1	3	2	1	1	2	1	2	1	2	1	2	1	?	1
S. crassifolius	1	2	1	1	2	1	1	1	2	1	2	1	2	2	2	2	2	2	2	2
S. dubius	1	1	1	1	2	1	2	1	1	1	1	1	1	2	1	1	1	1	2	2
S. ferrugineus	2/3	1/3	1/3	2	2	2	3	2	2	2	1	1	1	2/3	1	2	2	1	2	2
S. gyrocarpus	2/3	4	1	2	2	1	3	2	1	1	2	1	1	2	1	2	1	1	1	1
S. harmandii	1	2	1	1	1	1	1	1	2	1	1	1	1	1	2	2	2	1	2	2
S. hirsutus	1	2	1	1	1	1	3	2	1	1	1	1	1	2	1	2	4	1	2	1
S. latibractea	1	4	1	1	2	1	2/3	1/2	2	3	2	1	1	2	1	1	2/3	1	2	3
S. latistipulus	1	4	1	1	1/2	1	2	1	2	1	2	1	1	2	1	2	1	1	2	2
S. littoralis	2	1	1	1	2	1	2	1	1	1	1	1	1	2	1	2	2	1	2	2
S. macropterus	2	1	1	1	1/2	1	2	1	1	1	1	1	1/3	2	1	1	2	1	2	1/2
S. maingayi	1	1	1	1	2	1	2	1	1	1	1	1	3	2	1	2	2	1	2	1/2
S. merguensis	1	1/3	1	1	2	1	1	1	2	1	2	?	1	2	2	2	?	1	2	1
S. multiflorus	1	2	1	1	2	1	2	1	2	3	2	1	1	2	1/2	2	2	1	2	1
S. oblongifolius	3	1	3	2	2	1	2	1	1	1	2	1	2	1/2	2	1	2	1	2	2
S. parviflorus	2	1/3	1	2	2	2	3	2	1	1	2	1	1	3	1	2	1	2	2	1
S. persicinus	1	4	1	1	2	1	3	2	2	3	2	1	1/2	2	1	2	2	1	2	1
S. pottingeri	3	1/3/4	1	2	2	2	3	2	1	1	2	1	1/2	2	2	2	2	2	2	1
S. pulcher	3	3	1	2	2	1	3	1	1	1	1	1	2	2	2	2	2	2	2	1
S. purpureus	1	2	1	1	2	1	2	1	2	1	2	1	1	2	1	2	2	1	2	1
S. ridleyi	1	1	1	1	1	1	1	1	1	1	2	1	2	1	2	2	1	1	2	1
S. sanguineus	2	2	1	2	2	1	3	2	1	1	2	1	1	2	2	2	1	1	1/2	1
S. suberectus	1	4	1	2	2	1	3	3	1	1	1	1	2	2	1	2	1	1	2	2
S. viridis	1	4	1	1	1	1	2	1/2	2	3	2	?	1	2	1	1	?	1	2	2
B. monosperma	3	3	3	2	2	2	3	3	1	2	1	1	1	1	1/2	2	2	1	2	1
B. superba	3	3	1/3	2	2	2	3	3	1	2	1	1	1	1	1/2	2	2	1	2	1/2
M. buteiformis	2	1	1/3	2	2	2	3	3	1	2	1	2	1	1	1	1	?	1	2	1
M. pellita	2	1	1/3	2	2	1	3	3	1	2	1	2	2	1	1	1	2	1	2	1
Kunstleria	1	2	1	1	1	1	2	1/2	1	1	1	1	1	2	2	2	2	1	1	1

	41	42	43	44	45	46	47	48	49	50	51	52	53	54	55	56	57	58	59	60
S. acuminatus	2	1	1	4	2	1	2	1/2	1	2	1	2	2	2	2	2	2	1	2	2
S. albus	1	1	3	1	2	1	2	2	1	1	2	2	2	2	2	1	2	1	2	2
S. apoensis	1	2	3	1	2	1	2	1	1	2	2	2	2	2	2	1/2	2	2	2	2
S. auricomus	2	1	2	1	2	1	2	2	2	2	2	2	2	2	2	2	2	2	2	2
S. auritus	2	1	3	2	2	1	2	2	1	2	1	2	2	2	2	1	2	1	2	2
S. bracteolatus	2	1	1	1	2	1	2	2	?	?	?	?	?	?	?	?	?	?	?	?
S. crassifolius	2	1	1	1	2	1	2	2	1	2	1	2	2	2	2	2	2	1	2	2
S. dubius	2	1	2	1	2	1	2	2	1	2	1	2	2	2	2	1	2	2	2	2
S. ferrugineus	2	1/2	2	1	2	1	2	1	1	1	1	2	2	2	1/2	1	2	1	2	2
S. gyrocarpus	2	1	2	1	2	1	2	1	2	2	1	2	2	2	2	2	2	1	2	2
S. harmandii	2	1	2	3	2	1	2	2	1	2	1	2	2	2	2	2	2	1	2	2
S. hirsutus	2	1	3	1	2	1	2	1	1	2	1	1	2	2	1	1	2	1	2	2
S. latibractea	2	1	2	1	2	1	2	2	2	2	1	1	2	1	1	2	2	1	1/2	2
S. latistipulus	1	2	3	2	2	1	2	2	1	2	1	2	2	2	2	1/2	2	2	2	2
S. littoralis	2	2	3	1	2	1	2	1	1	2	1	2	2	2	2	2	2	2	2	2
S. macropterus	2	1/2	2	1	2	1	2	1	1	2	1	1	2	1	1	1/2	2	1	2	2
S. maingayi	1	2	3	2	2	1	2	1	1	2	2	2	2	2	2	1	2	2	2	2
S. merguensis	2	1	1	3	2	1	2	2	2	2	1	2	2	2	2	2	2	1	2	2
S. multiflorus	2	1	2	1	2	1	2	1/2	1	1	1	2	2	2	2	1/2	2	1	2	2
S. oblongifolius	2	1	2	3	2	1	2	1/2	2	2	1	2	2	2	2	1/2	2	1	2	2
S. parviflorus	2	2	1	1	2	2	1/2	2	1	2	1	2	2	2	2	1	2	1	2	2
S. persicinus	2	2	3	2	2	1	2	2	1	2	2	2	2	2	2	1	2	2	2	2
S. pottingeri	2	1	2	3	2	1	2	1	1	2	1	2	2	1	1	1	2	1	2	2
S. pulcher	3	2	1/2	1	2	1	2	1	1	1	1	2	2	2	2	1	2	1	2	2
S. purpureus	2	1	1	1	2	1	2	1	2	2	1	2	2	2	1	2	2	1	2	2
S. ridleyi	2	2	2	4	2	1	1/2	1	2	2	1	2	2	2	1	2	2	1	2	2
S. sanguineus	2	1	2	1	2	1	2	2	1	2	1	2	2	2	2	1	2	1	2	2
S. suberectus	2	1/2	2	3	2	1	2	1/2	1	2	1	2	2	2	2	2	2	1	2	2
S. viridis	2	2	2	1	2	1	2	1	1/2	1	1	1/2	2	2	1/2	2	2	1	2	2
B. monosperma	2	1/2	2	1	1	2	2	1	1	2	1	1	1	1	1	1	2	1	1	2
B. superba	2	1/2	2/3	1	1	2	2	1	1	2	1	1	1	1	1	2	2	1	1	1
M. buteiformis	2	1	3	1	1	2	1	2	1	2	1	1	1	2	1	2	2	1	1	1
M. pellita	3	2	2	3	1	2	1	2	1	2	1	1	2	2	1	2	2	1	1	2
Kunstleria	2	1/2	2	3	2	2	2	2	1	2	1	1	2	1	1	1	2	1	1	2

(Table 2.1 contd)	61	62	63	64	65	66	67	68	69	70	71	72	73	74	75	76	77	78	79	80
S. acuminatus	2	2	1	1	3	1/2	1	1	1	1/2	1	2	1	2	1	1/2	2	1	1	1
S. albus	2	2	1	1	3	1	1	1	1	1	2	3	1	2	1	2	2	1	2	1
S. apoensis	2	2	1	1	1	1	2	1	1	1	2	3	1	2	1	1	?	?	?	?
S. auricomus	2	2	1	1	1	1	1	3	1	1	3	2	1	2	1	2	?	?	?	?
S. auritus	2	2	1	1	3	1	2	1	1	1	1	2	1	2	2	1	?	?	?	?
S. bracteolatus	?	?	?	1	?	1	1	3	4	1	2	1	1	2	1	3	?	?	?	?
S. crassifolius	2	2	1	1	1	1	1	1	4	2	1	2	1	2	2	2	?	?	?	?
S. dubius	2	2	1	1	3	1	1	1	1	1	1	2	1	2	2	1	1	3	2	1
S. ferrugineus	2	2	1	1	1	1	1	3	1	1	1	1	1	2	2	1	2	2	2	2
S. gyrocarpus	2	2	1	1	1	1	1	1	1	1	3	2	1	2	1	1	2	2	2	2
S. harmandii	2	2	1	1	1	1	1	1	1	1	2	2	1	2	2	1	1	2	2	2
S. hirsutus	2	2	1	1	2/3	1	1	1	1	1/2	1	1	1	1	2	1/2	1/2	2	2	1
S. latibractea	1	2	1	1	3	1	2	1	1	1	1/2	1	1	2	1	1	?	?	?	?
S. latistipulus	2	2	1	1	1	1	2	1	1	1	2	3	1	2	2	2	2	3	1	1
S. littoralis	2	2	1	1	1	1	2	1	1	1	2	1	1	2	2	2	2	2	1	1
S. macropterus	2	2	1	1	2/3	1	1	1	1	2	2	1	1	1/2	2	2/3	1/2	1	2	1
S. maingayi	2	2	1	1	1	1	2	1	1	1	1	1	1	2	2	2	2	3	1	1
S. merguensis	2	2	1	1	1	1	1	1	4	1	1	1	1	2	2	2	?	?	?	?
S. multiflorus	2	2	1	1	1	1	1	1	1	1	1	2	1	2	1	2	2	3	2	2
S. oblongifolius	2	2	1	1	3	1	1	1	1	1	1	2	1	1	1	2	2	3	2	1
S. parviflorus	2	2	1	1	3	1	1	1	1	1	1	2	1	1	2	1	1	2	2	2
S. persicinus	2	2	1	1	1	1	2	1	1	1	2	3	1	1	1	2	?	?	?	?
S. pottingeri	2	2	1	1	3	1	1	3	4	1	1	1	1	2	1	2	2	2	2	2
S. pulcher	2	2	1	1	3	1	1	1	2	2	2	2	2	2	2	2	2	3	2	2
S. purpureus	2	2	1	1	2	1	1	1	2	2	1	2	1	2	2	1	2	3	1	1
S. ridleyi	2	2	1	1	3	1	1	1	1	2	1	1	1	1	2	1	2	2	2	1
S. sanguineus	2	2	1	1	1	1	1	3	1	1	3	2	1	2	1	2	2	3	2	2
S. suberectus	2	2	1	1	3	1	1	1	3	1	3	2	1	1	2	1	1	2	2	2
S. viridis	2	2	1	1	3	1	1	1	1	1	2	1	1	2	2	2	?	?	?	?
B. monosperma	1	1	1	1	2	2	3	1	4	1	2	2	3	2	2	3	1	2	2	2
B. superba	1	1	1	1	2	2	4	1	4	1	2	2	3	2	2	3	1	2	2	2
M. buteiformis	2	1	1	1	3	2	3	1	4	1	1	3	1	2	1	3	2	2	1	1
M. pellita	2	1	1	1	3	2	3	1	4	1	2	3	1	2	1	3	2	2	1	2
Kunstleria	1	1	1	2	3	1	1	2	4	1	2	2	1	1	2	2	2	2	1	2

	81	82	83	84	85	86	87	88	89	90	91	92	93	94	95	96	97
S. acuminatus	1/2	2	1	3	1	2	2	1	2	2	1	2	1	1	1	2	1
S. albus	2	2	1	3	1	1/2	2	1	2	2	1	2	2	1	1	2/3	1
S. apoensis	2	2	1	3	1	1/2	2	1/2	2	2	2	1	1	1/2	1	2/3	2
S. auricomus	1/2	2	1	2/3	1	2	2	1	1	2	1	1	?	1	?	2/3	2
S. auritus	?	?	?	?	?	?	?	?	?	?	1	1	?	2	?	4	2
S. bracteolatus	?	?	?	?	?	?	?	?	?	?	?	?	?	?	?	?	?
S. crassifolius	1/2	2	1	3	1	1	2	1/2	2	1	1	2	?	1	?	3	1
S. dubius	1/2	2	1	3	1	1	2	1	1	2	1	2	?	1	?	3	2
S. ferrugineus	1/2	2	1	2/3	2	1/2	1/2	2	3	1	1	1	2	1	1	1	1
S. gyrocarpus	2	2	1	3	1/2	1	1/2	2	2	2	1	1	?	1	1	?	3
S. harmandii	2	2	1	4	1	2	1/2	1/2	2/3	2	1	3	?	1	1	2/3	1
S. hirsutus	2	2	1	3	1	2	2	1/2	3	1	1	1	2	1	1	1/2	2
S. latibractea	1/2	2	1	3/4	1	2	2	1	3	2	1/2	2	1	1	1	2/4	1
S. latistipulus	2	2	1	3/4	1	1/2	2	1	3	2	1/2	1	?	1/2	?	3	2
S. littoralis	2	2	1	3	1	2	2	1	3	2	1/2	2	2	1/2	1	2/3	1
S. macropterus	2	2	1	3	1	2	2	1/2	3	2	1	1	2	1	1	1/2	2
S. maingayi	2	2	1	3	1	2	2	1/2	2	2	1/2/3	2	1	1/2	1	3	2
S. merguensis	?	?	?	?	?	?	?	?	?	?	1	2	?	1	1	3/4	2
S. multiflorus	1/2	2	1	1/4	1	1	2	1	1	2	1	1	2	1	1	1/2/4	2
S. oblongifolius	1/2	2	1	3	1	2	2	1	1/3	1	1	2	2	1	1	2/3	1
S. parviflorus	1	1	3	3	3	1	2	2	1	2	1	1	3	1	2	2/3	2
S. persicinus	1/2	2	1	3	1	2	2	2	2	2	1/2	3	1	1	1	2/3	1
S. pottingeri	1/2	2	1	3	1/2	1	2	1	3	2	1	1	?	1	1	2/3/4	2
S. pulcher	1	1	1	2	1	1	2	2	?	2	1	1	?	1	2	?	3
S. purpureus	2	?	1	3	1/2	1	2	2	2	1	1	2	2	1	1	2/3	1
S. ridleyi	1/2	2	1	3	1	2	2	2	2	2	1	1	?	1	1	?	2
S. sanguineus	1/2	2	1	3	1	2	1/2	2	2	2	1	1	1	1	1	1/2	3
S. suberectus	?	?	1	3	1	1	2	2	3	2	1	1	2	1	1	2/3/4	2
S. viridis	1/2	2	1	4	1	2	2	1	3	2	1	3	2	1	1	2/3	1
B. monosperma	1	1	1	2	1/2	1	2	2	1	1	1/2	1	2	2	2	1/2	2
B. superba	1	1	1	?	1	1	2	1/2	1	1	1/2	1	?	2	2	1/2	2
M. buteiformis	1	1	1	2	1	1	2	2	1	2	1/2/3	1	3	2	2	2/3	3
M. pellita	?	?	?	?	?	?	?	?	?	?	2/3	1	3	2	2	2/3	3
Kunstleria	1	1/2	2	3	1/2	1	1	1/2	1	2	1	1	1	1	1	1/2	3

Table 2.2. Quantitative characters used for the phylogenetic analysis.

Characters: / Species:	stipule length by width [12]	angle of nerves [29]	pairs of nerves [31]	pedicel length in mm [39]	connate filaments [65]	anther size in mm [67]	A/E [97]			foot layer in µm	endexine in µm
S. acuminatus	6–7 × 1–1.5	(45–)50–60	6–9	2–5	> 2/3	0.2–0.5	0.22	(0.26)	0.30	0.18–0.32	0.13–0.23
S. albus	2 × 1	35–50	3–11	1–1.5	2/3	ca.0.5	0.15	(0.27)	0.31	0.046–0.13	0.27–0.32
S. apoensis	7 × 3	40–55	7	2–3	1/2	0.5–1	0.23	(0.28)	0.38	0.2–0.4	0.2–0.4
S. auricomus	? × 4	less than 40	5–7	1–2	1/2	0.3	0.23	(0.27)	0.41	?	?
S. auritus	7 × 2–3	45	12–16	1–2	2/3	0.5	0.25	(0.33)	0.45	?	?
S. bracteolatus	?	35–40	8–10	?	?	?	?			?	?
S. crassifolius	5–8 × 2	50	7–10	3	1/2	0.3	0.16	(0.21)	0.29	?	?
S. dubius	8 × 2	32–38	4–7	1.5–2.5	2/3	0.3	0.22	(0.31)	0.36	?	?
S. ferrugineus	3–6 × 2–6	45–60	4–8	1–4	< 1/2	0.2	0.17	(0.22)	0.26	?	?
S. gyrocarpus	?	40	7–11	1	1/2	0.2	0.4	(0.44)	0.49	0.6–0.13	0.3–0.4
S. harmandii	1.5–2 × 0.5–1.5	50	5–10	2–6	1/2	0.3	0.18	(0.2)	0.24	?	?
S. hirsutus	7–17 × 2–4	15–30(–40)	5–7(–11)	1–1.5	3/4	0.2–0.3	0.31	(0.37)	0.43	?	?
S. latibractea	13–18 × 5–8	50–65	9–11	1–2	1/2–2/3	0.3–0.75	0.13	(0.18)	0.23	0.18–0.27	0.5–0.69
S. latistipulus	11–12 × 7–12	45–55	7–12	1–2(–4)	1/4–1/2	0.75	0.2	(0.3)	0.4	0.32–0.6	0.55–0.69
S. littoralis	4–8 × 1.5–4	20–45	5–8	1.5–3	< 1/2	0.5	0.2	(0.24)	0.29	?	?
S. macropterus	7 × 1.5	35–40(–50)	4–8	1–1.5	> 3/4	0.3	0.24	(0.36)	0.39	0.27	0.18–0.4
S. maingayi	3–5(–10) × 2–3	35–45	5–8	1–2	1/3–1/2	0.5–0.75	0.27	(0.31)	0.36	0.06–0.13	0.33–0.46
S. merguensis	5–6 × 1–2	55	10	1	1/2	0.3	0.27	(0.38)	0.44	0.2–0.3	0.2–0.4
S. multiflorus	10–16 × 3–5	50	9–14	1–2	up to 1/2	0.2–0.3	0.3	(0.33)	0.37	?	?
S. oblongifolius	?	20–40	8–10	1	2/3	0.2	0.16	(0.21)	0.25	0.23–0.32	0.37–0.69
S. parviflorus	4–6 × 5–7	40	7–9	2–4	2/3	0.3	0.27	(0.3)	0.33	0.093	0.37–0.5
S. persicinus	7–12 × 3–5	50–60	7–16	1–2	1/2	0.5–0.75	0.24	(0.26)	0.3	0.055(–0.11)	0.55–0.8
S. pottingeri	8–14 × 2–5 or 4 × 6	40–45	7–12	1–3	1/2–2/3	0.2–0.5	0.25	(0.34)	0.41	0.2–0.4	0.26–0.4
S. pulcher	?	40–50	4–7	2	2/3	0.3	0.5	(0.58)	0.65	?	?
S. purpureus	?	60–80	6–11	2	3/4	0.3	0.12			?	?
S. ridleyi	?	40–45	7–9	1.5–2	up to 3/4	0.2	0.38			?	?
S. sanguineus	7–12 × 4–7	30–45	6–11	0.25–1.5	< 1/2	0.2	0.43	(0.44)	0.47	0.093–0.13	0.27–0.37
S. suberectus	7–10 × 2–4	40–50	6–8	1–2	2/3	0.2–0.3	0.29	(0.33)	0.37	?	?
S. viridis	11–13 × 4–5	50–55	9–12	1–2	1/2–3/4	0.3–0.5	0.2	(0.25)	0.31	0.18–0.27	0.23–0.37
B. monosperma	5–7 × 1.5–4	35	5–9	1.5–4	> 5/6	1–2	0.30	(0.36)	0.44	0.093–0.18	0.5–0.64
B. superba	5–7 × 1.5–6	30–45	6–8	2–4.5	nearly all	2.5–4.5	0.36	(0.43)	0.51	0.13–0.26	0.33–0.66
M. buteiformis	8–18 × 5–14	35–50	5–9	5–15	2/3	1.5	0.43	(0.5)	0.57	?	?
M. pellita	15–30 × 10–16	25–40	5–8	7–10	> 2/3	1–1.5	0.53	(0.62)	0.67	0.06–0.13	0.26–0.53
Kunstleria	6 × 2.5–5	25–55	4–8(–12)	up to 1	1/2–3/4	0.3–0.5	0.35	(0.39)	0.45	0.03–0.24	0.1–0.7

Stem

Most species in this analysis are lianas, only *Butea monosperma* is a tree, and the two species of *Meizotropis* are shrubs. The stem characters are all taken from branches with a diameter of 1 cm or less.

In some species of *Spatholobus* the stem is hollow [1]. Small ants, of the genus *Cladomyrma*, may inhabit the stem as is often observed in *S. oblongifolius* (Moog & Maschwitz, 1994). The ants enter the stem by small holes. Hollow branches occur also in *S. auritus, S. bracteolatus, S. hirsutus,* and *S. pottingeri. Spatholobus ferrugineus* and *S. macropterus* show both states. Moog found ants as well in specimens of most probably *S. bracteolatus.* The other species of *Spatholobus* do not have hollow branches, or, if so, only in part.

Stems are often glabrescent, the indumentum [2] being usually present on the younger branches only. Most species in *Spatholobus* have either nearly glabrous stems, or a pubescent/puberulous or sericeous indumentum. In contrast to these *S. bracteolatus, S. pottingeri,* and *S. pulcher* are hirsute or strigose. The genera *Butea* and *Meizotropis* are usually densely pubescent, except *M. pellita*, which is covered by a dense pilose type of indumentum. *Kunstleria* is pubescent as well, except for one species, *K. forbesii*, which has a woolly indumentum on the branches. This character state is unique within *Kunstleria* and very rare in the tribe Millettieae. Therefore it is not included in the data set for the genus as a whole.

Lenticels [3] are small and rounded in most species of *Spatholobus* as well as in those of the genera *Butea* and *Meizotropis*. In *Kunstleria* and some species of *Spatholobus* (*S. acuminatus, S. albus, S. suberectus, S. viridis*) the lenticels are wart-like. Elongated lenticels (Fig. 2.1a) are found in *S. dubius, S. latibractea, S. littoralis,* and *S. pulcher*. The two last types of lenticels are more conspicuous than the first.

The exudate [4] is a specific feature and visible as a dark stained residue on the branches where they are cut off. Also on some of the field labels this feature is mentioned. The species of *Butea* and *Meizotropis* all have exudate. Although the branches of the species of *Kunstleria* also contain a dark substance, they do not show the dark stains of an exudate. The only species that is different in this character is *K. forbesii*, which most probably produces exudate. Within *Spatholobus* there are species with and without exudate. The only polymorphic species are *S. littoralis* and *S. pottingeri*.

The bark of the branches [5] may be either smooth or with ridges or wrinkles. Most species of *Spatholobus* have a smooth bark. The other species of *Spatholobus* and the other genera in this study usually have small ridges or wrinkles on the bark (Fig. 2.1b). Two species, *S. auritus* and *S. littoralis*, show a polymorphism here.

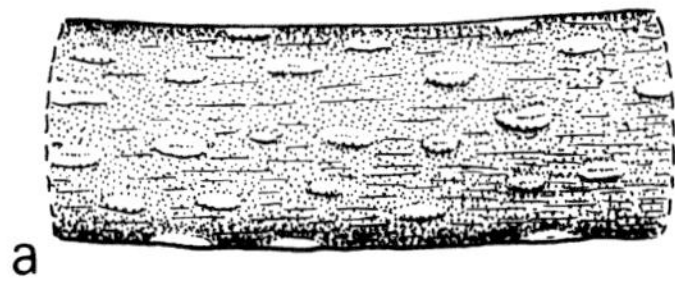

Figure 2.1. a. Part of the stem of *Spatholobus latibractea* showing lenticels. b. Part of the stem of *S. latistipulus* showing small ridges on the stem. — After a drawing by J. Wessendorp in Ridder-Numan (1992).

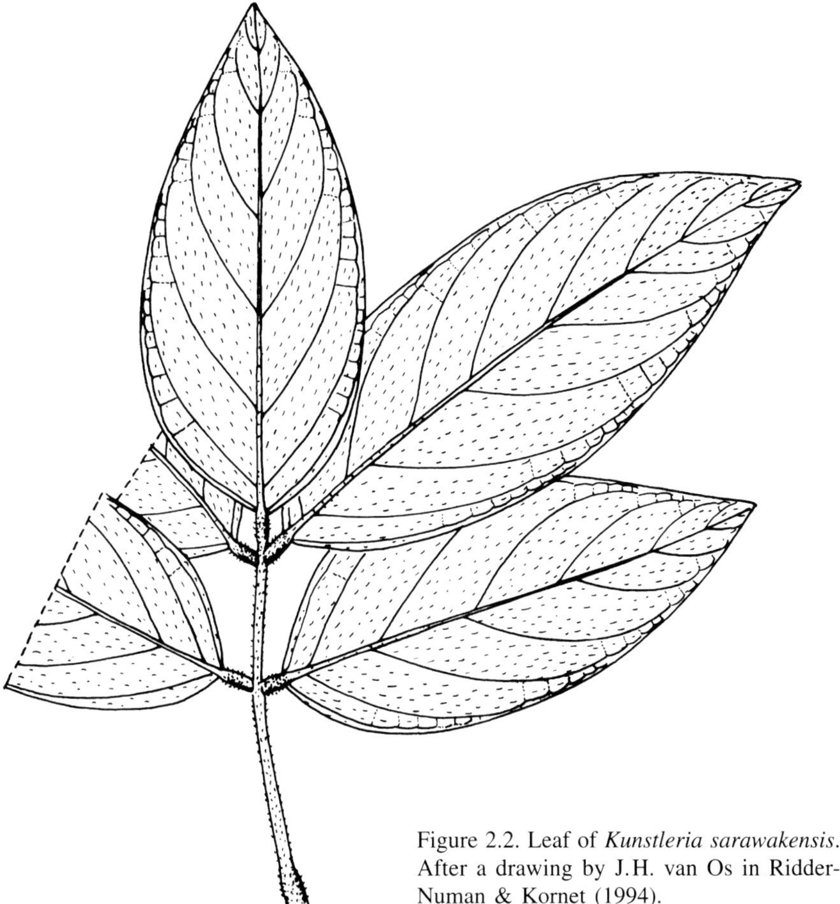

Figure 2.2. Leaf of *Kunstleria sarawakensis*. After a drawing by J.H. van Os in Ridder-Numan & Kornet (1994).

Leaves

The genera *Butea, Meizotropis*, and *Spatholobus* have trifoliolate leaves (Fig. 2.3, 2.4, 2.7, 2.8, 2.10). This is a common feature in the tribe Phaseoleae. The lateral leaflets in this tribe usually have an enlarged basiscopic side (Geesink, 1984). All leaves have stipules and stipellae, although these may be early caducous. The absence of stipules in the available material is responsible for the question marks in the data matrix for stipule characters [7–12]. The genus *Kunstleria* has species with uni- and trifoliolate leaves, but most of the species have more than five leaflets (Fig. 2.2). Here the leaflets are (almost) symmetrical. The leaves of *Kunstleria* have stipules, but are exstipellate.

The ultrajugal part of the rachis [6] is very short or absent in some species of *Spatholobus*. The absence of the last part of the rachis in combination with the narrowly obovate shape of the leaflets is very characteristic for *S. oblongifolius*(Fig. 2.4). Sometimes, perhaps due to the absence of the ultrajugal part, there are four or five leaflets per leaf in contrast to the usual three. Other species with short or nearly absent ultrajugal parts are *S. merguensis* and *S. multiflorus*. One species, *S. albus*, has an ultrajugal

Figure 2.3. Leaf of *Spatholobus littoralis*.

part varying from nearly 0 up to 1/3 of the total length of the rachis, and is thus coded polymorphic for this character.

In most species the stipules [7] are straight (Fig. 2.5a). In some, however, the stipules curl back (Fig. 2.5b). These stipules seem to be thicker than the stipules that are straight. Recurving stipules are present in the genus *Butea* and some of the species of *Spatholobus*: *S. albus, S. ferrugineus, S. parviflorus* and *S. sanguineus*. It was not possible to code this character for all species, because in some species all stipules were caducous.

Stipules [8] are easily caducous in most species, as already mentioned above; in others they are more persistent, e.g., in *Spatholobus auritus, S. hirsutus, S. latibractea, S. latistipulus,* and *S. multiflorus*.

Figure 2.4. Leaf of *Spatholobus oblongifolius*. Shown are the top leaflet and the rachis. Note the absence of the ultrajugal part of the rachis.

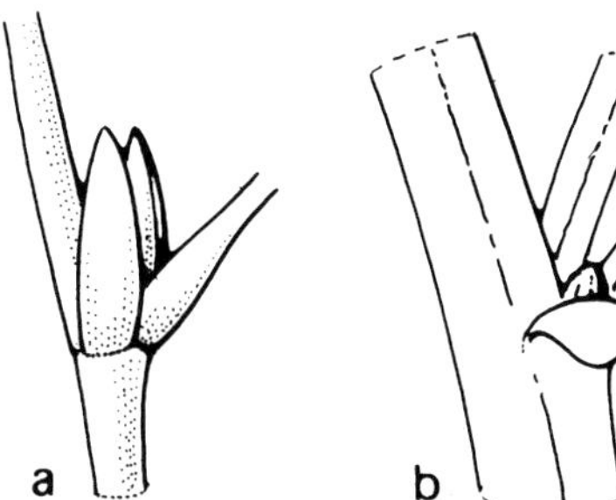

Figure 2.5. a. Straight stipules (*S. auritus*), b. curled stipules (*S. ferrugineus*).

Stipules [9] are either more or less symmetrical or not. Most species, as far as is known, have more or less symmetrical stipules. Only those of *Spatholobus latistipulus*, which are relatively wide, are asymmetric.

Sometimes the veins [10] of the stipules are prominent, in other cases the veins are not raised (and not visible). Two species have only 1–3 veins: *Spatholobus harmandii* and *S. maingayi*; others have many parallel veins.

Most species have an indumentum [11] on the stipules. Some species of the genus *Spatholobus*, however, are glabrous or glabrescent: *S. apoensis, S. hirsutus, S. latistipulus, S. littoralis, S. macropterus, S. maingayi* and *S. ridleyi.*

The length [12] of the stipules is usually two to three times the width. Some species have shorter stipules (about as long as they are wide): *Spatholobus albus, S. ferrugineus, S. latistipulus, S. parviflorus* and the two species of the genus *Meizotropis*. Others, *S. acuminatus, S. hirsutus* and *S. macropterus* have relatively long stipules that may be more than four times as long as they are wide. In Figure 2.6 the relative length of the stipules is given, calculated from the averages of the values of length and width of the stipules.

The colour [13] of the pulvinus of the leaflets can be darker than the rachis itself (Fig. 2.3). In most cases the colour is the same, but in some species of *Spatholobus* the pulvinus is nearly black in dried material. These species are: *S. apoensis, S. bracteolatus, S. hirsutus, S. latistipulus, S. macropterus, S. maingayi, S. oblongifolius* and *S. ridleyi.* The only species polymorphic for this character are *S. bracteolatus* and *S. pottingeri.* The material of the first species, however, is very scarce, and it may be possible that one of the sterile specimens studied is misidentified.

The pulvinus can be either glabrous (or sparsely hairy) or with an indumentum [14], which is either pubescent to hirsute or with longer hairs and more densely pubescent. The latter state is found in *Meizotropis pellita* and in several species of *Spatholobus.*

The stipellae of the leaflets are in most cases caducous [15], and are in most species only present in juvenile leaves. In other species they are more or less persistent and may even be present when the leaflets have dropped. The persistence of the stipules is not associated with this feature of the stipellae. In *Spatholobus latibractea, S. latistipulus* and *S. multiflorus* both stipules and stipellae are more or less persistent (Fig. 2.7), while in *S. auritus* and *S. hirsutus* the stipules stay longer on the leaves than the stipellae.

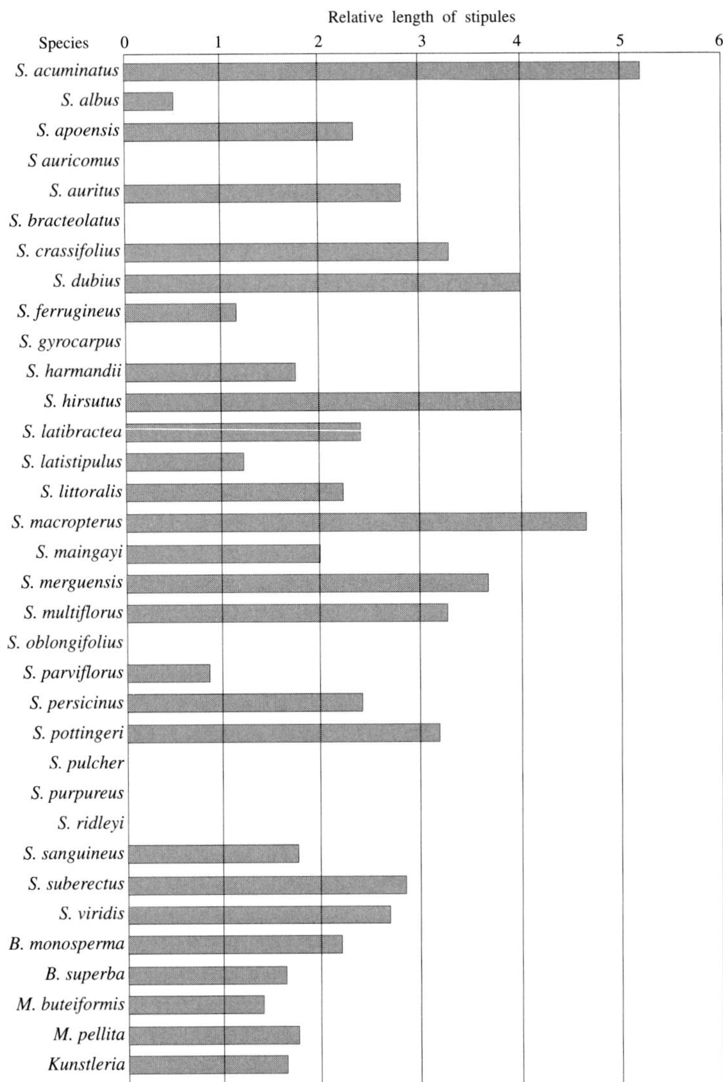

Figure 2.6. Relative length of the stipule.
Species of which the stipules were not seen are not represented by a bar.

Spatholobus dubius, S. harmandii, S . pottingeri and *S. ridleyi*, however, have persistent stipellae, but early caducous stipules. In the genus *Kunstleria* no stipellae are present,
and a question mark is entered for the two characters concerning stipellae [15 & 16].
The second character of the stipellae that I used in the analysis is the length [16].
Stipellae are usually not longer than the petiolule (Fig. 2.3). In some species of *Spatholobus*, however, the stipellae are conspicuously longer than the petiolule. It was not
always easy to score this character, because some of the more persistent and long sti-

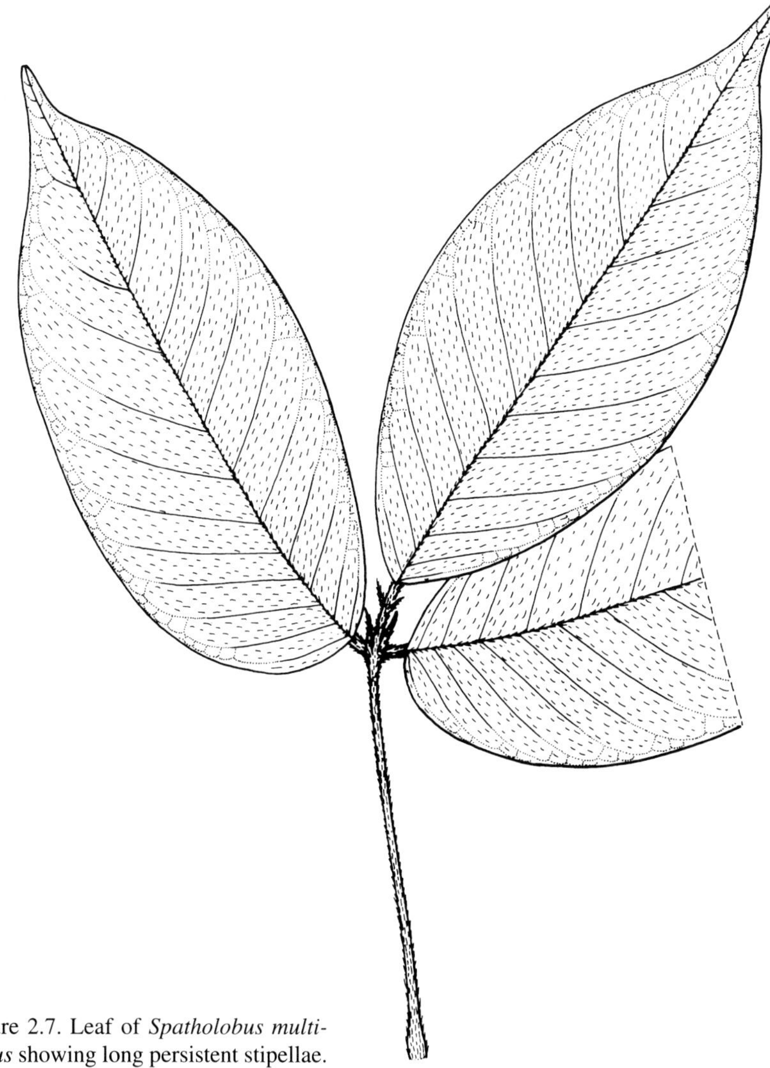

Figure 2.7. Leaf of *Spatholobus multi-florus* showing long persistent stipellae.

pellae had a tendency to break off at the tip. There are only two species with such long stipellae that are persistent, *S. multiflorus* (Fig. 2.7) and *S. pottingeri*. In all other cases the stipellae are caducous.

The upper surface [17] of the leaflets in some species of *Spatholobus* deviated from the usual green/brownish (dried)colour. In these species the upper surface is glaucous.

The lower surface [18] may be dull brown versus the usual green/brownish colour of dried material. The character is not in all cases associated with a glaucous upper surface.

All species of *Spatholobus* and *Kunstleria* have leaves with a glabrous upper surface [19]. In the genera *Butea* and *Meizotropis* the leaves are hairy above. *Meizotropis pellita* is the most densely hairy species, having a hirsute to strigose indumentum on the upper surface of the leaflets. When the leaves are very young they are covered with a thick layer of hairs on both sides. Also other species, e.g. *Spatholobus ferrugineus*, have an indumentum on the upper surface of the very young leaves, but this disappears as soon as the leaves expand. In microscopical sections of the leaf surface the bases of the hairs are often still visible.

The lower surface indumentum [20] is more variable, and although many species of *Spatholobus* are glabrous (glabrescent) on the lower leaf surface, about half of the species show one or more types of pubescence. Three species of *Spatholobus* are polymorphic for the indumentum on the lower surface. In two of these species infraspecific taxa have been recognised mainly on account of this character. In the revision of *Spatholobus* (Ridder-Numan & Wiriadinata, 1985) infraspecific taxa are only recognised in *S. ferrugineus*. Apart from the indumentum, the shape of the leaflets is also aberrant in these taxa. Baker (1867) also recognised a separate variety based on the glabrous leaflets in *S. parviflorus* (i.e., var. *denudatus*).

To represent the shape of the leaflets the terminal leaflet [21] is taken to avoid confusion between asymmetric and symmetric leaflets. Most species of *Spatholobus* and *Kunstleria* (except *K. forbesii*) have only elliptic leaflets (Fig. 2.2, 2.3), but some species have obovate (Fig. 2.4) or ovate (Fig. 2.7) leaflets as well. *Spatholobus ferrugineus* and *S. gyrocarpus* are polymorphic for this character: their leaflets can be elliptic, ovate, or obovate, but never are only elliptic. The state 'elliptic' was scored if only elliptic leaflets were present (i.e. obovate or ovate leaflets absent). When the outgroup *Kunstleria* is considered as possessing the plesiomorphic state, the state 'elliptic' is plesiomorphic. The coding for this character is in fact favouring the apomorphic situation in polymorphic species. I chose this way of coding to avoid too many polymorphic situations.

The apex of the terminal leaflet [22] is in most cases acute or acuminate to cuspidate (Fig. 2.3, 2.4). In *Butea* the leaflets are usually emarginate at the apex (Fig. 2.8). In some species of *Spatholobus* emarginate apices are present, usually in obovate leaflets; sometimes in combination with leaflets which have acute or abruptly acuminate apices. An apex that is abruptly acuminate (Fig. 2.7) is found in all types of leaf shapes.

The base of the leaflet [23] is usually acute to obtuse or more rounded (Fig. 2.3) to subcordate. In some species the base of the leaflet can be decurrent (Fig. 2.4 and 2.8) in the species of, e.g., *Butea* and *Meizotropis, Spatholobus oblongifolius* and sometimes in *S. ferrugineus*. The latter species has a rather variable leaf shape.

Symmetrical lateral leaflets [24] are present in the species of *Kunstleria* and in more than half of the species of *Spatholobus* (Fig. 2.2, 2.3). It is obvious that one of the characteristic features for the tribe Phaseoleae, strongly asymmetrical lateral leaflets, does not hold for *Spatholobus*. In the genera *Butea* and *Meizotropis*, and the rest of *Spatholobus* asymmetric leaflets occur as may be expected in Phaseoleae (Fig. 2.8, 2.10).

Figure 2.8. Leaf of *Butea monosperma*. Note the emarginate apex, the decurrent base, and the strongly asymmetric lateral leaflets.

The main nerve [25] can be either flat, raised, raised in a furrow, or sunken. Some species never have the main nerve sunken, others have the main nerve usually sunken, e.g. *Kunstleria*. This character was checked with leaf anatomical sections. The main nerve is never sunken in species with strongly asymmetrical leaves.

In contrast with the main nerve, the secondary nerves [26] are usually not raised on the upper surface. Only in some species with relatively large leaves, i.e. the species of *Butea, Meizotropis buteiformis, Spatholobus ferrugineus, S. parviflorus* and *S. pottingeri,*

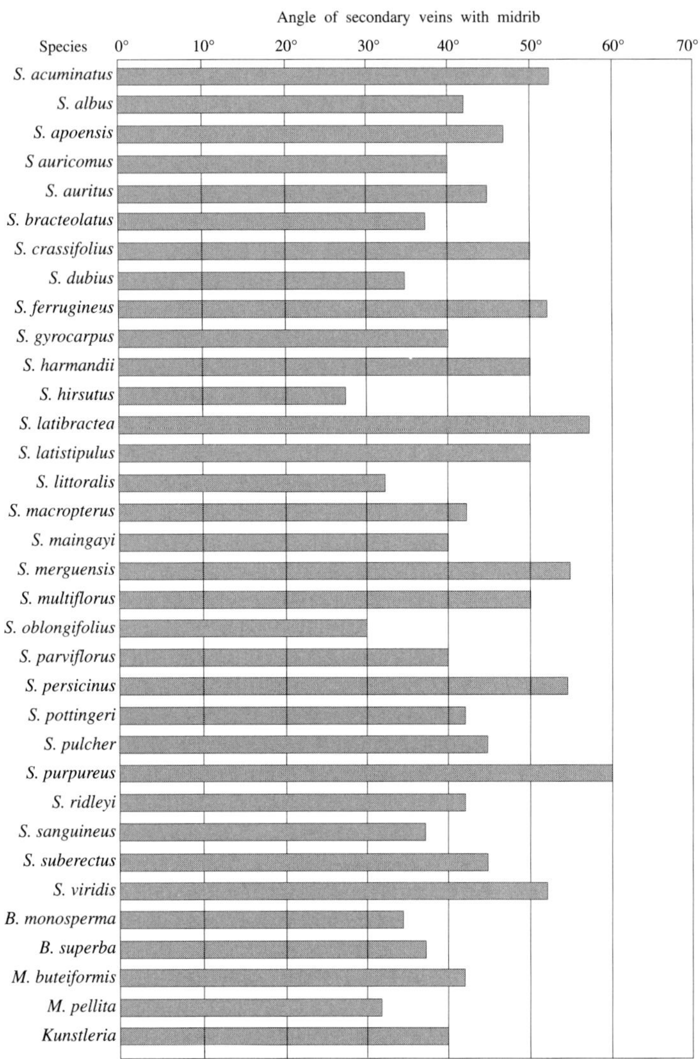

Figure 2.9. The average values of the angles between midrib and secondary veins.

both main and secondary nerves are raised. However, not in all species with large leaves both main and secondary nerves are raised at the upper surface.

On the lower surface [27] the veins may be prominent or not. In some species of *Spatholobus* (*S. acuminatus, S. albus, S. crassifolius, S. harmandii, S. merguensis, S. ridleyi*) only the midrib is prominent. In other species of *Spatholobus* and all species of *Kunstleria* both the main nerve and first order veins are prominent (Fig. 2.2, 2.3). The species with larger leaflets, *Butea, Meizotropis* and some species of *Spatholobus*, also have a prominent venation. *Spatholobus latibractea* may have leaflets with the venation prominent or not.

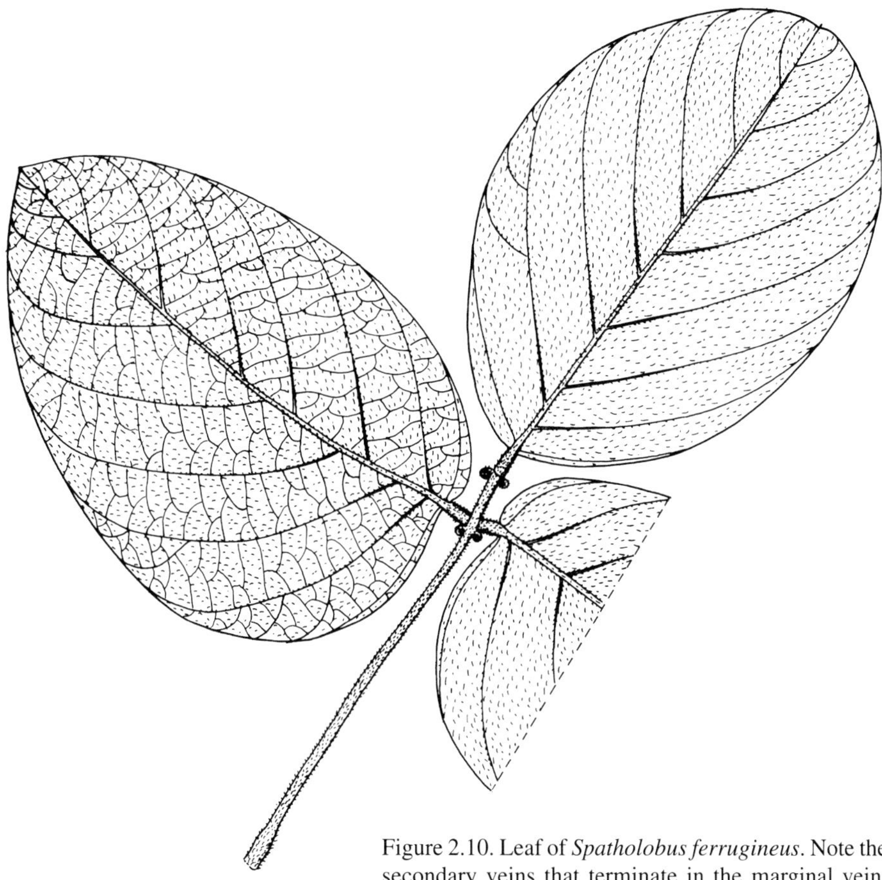

Figure 2.10. Leaf of *Spatholobus ferrugineus*. Note the secondary veins that terminate in the marginal vein.

The venation [28] is usually reticulate (Fig. 2.10) or reticulate-scalariform. In *Spatholobus latibractea, S. viridis* and *Kunstleria* the venation can have both states. A scalariform venation is present in the genera *Butea* (Fig. 2.8) and *Meizotropis*, and in *S. suberectus*.

The secondary veins are curved towards the top. The angle of the secondary nerves [29] with the midrib is taken at the place of the secondary vein in the middle of the lamina. In this way there is no difference between the course of a geniculate and a curved type of secondary veins. Most species have a low or moderately angled course. Some *Spatholobus* species, however, have high angles (more than 50°), others have low angles with the midrib (less than 40°). The other species have values in between, but the average of these values can be divided into two groups: one more than 45°, the other less than 45° (Fig. 2.9).

Secondary nerves [30] in most species of *Spatholobus* and *Kunstleria* terminate diffusely towards the margin, and do not form marginal arches (Fig. 2.3, 2.4). In *S. ferrugineus* and in the species of *Butea* and *Meizotropis* the nerves terminate in the margin (Fig. 2.8, 2.10). In *S. auritus, S. latibractea, S. multiflorus, S. persicinus*, and

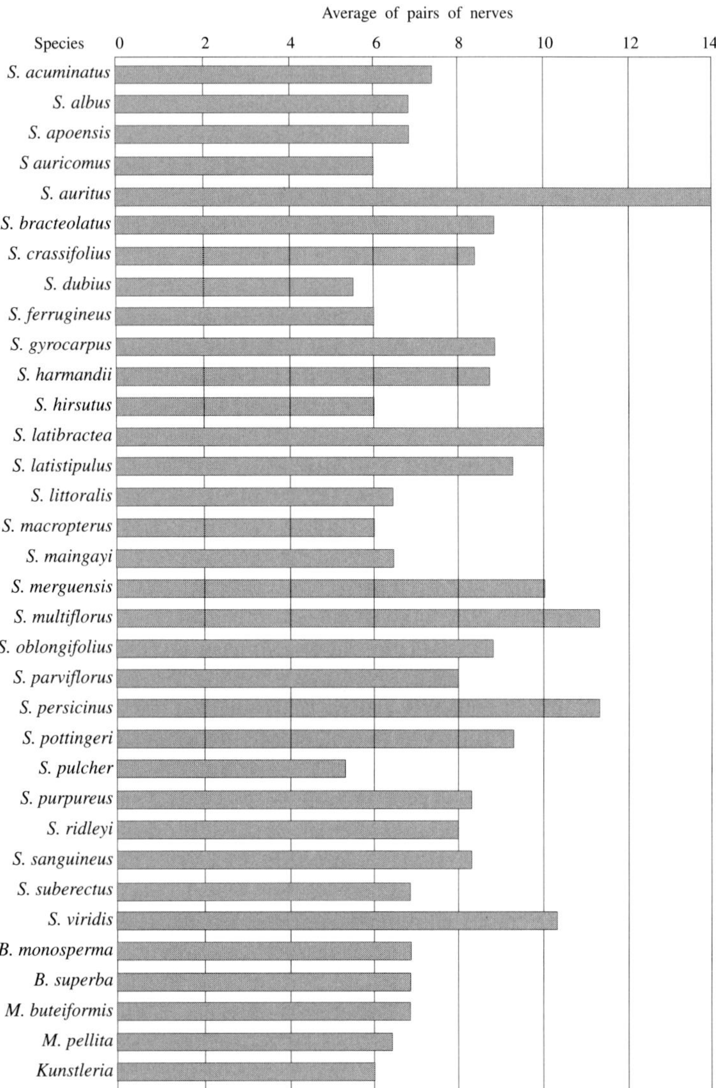

Figure 2.11. Average of the number of pairs of secondary nerves.

S. viridis the secondary veins do not end in the margin but in the vein above and form marginal arches (Fig. 2.7).

The number of pairs of secondary nerves [31] in most species of *Spatholobus* and in all species of the other genera studied is five to eight. Some species have more than eight pairs of secondary veins, and *S. auritus* and *S. multiflorus* always have more than 10 pairs of veins. The average of the number of pairs of nerves was taken to code this character (Fig. 2.11).

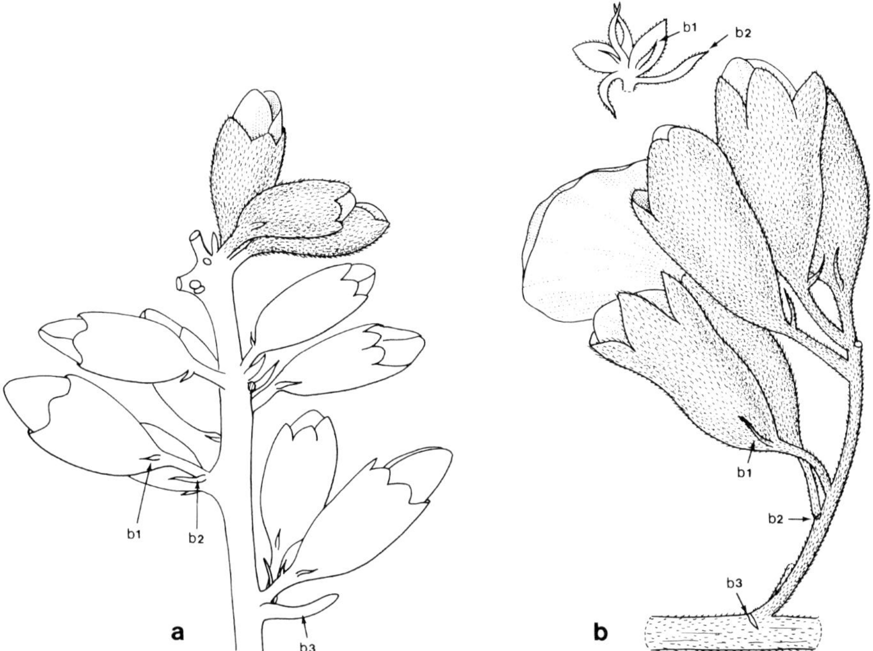

Figure 2.12. a. Part of the inflorescence of *Spatholobus suberectus* showing brachyblasts. b. Part of the inflorescence of *S. pottingeri*; b1 = bracteole, b2 = bract of the flower, b3 = bract of the side branch. Not all bracts are visible. After Ridder-Numan & Wiriadinata (1985).

Inflorescence

The inflorescence is variable among the species. Mostly there is a main axis with secondary branches bearing the (single) flowers directly or on an extra order of side branches (tertiary branches). Each flower is subtended by two bracteoles on the pedicel and a bract. Instead of bearing a single flower it is also possible that the side branch is reduced to a brachyblast (Fig. 2.12a). These brachyblasts are in some cases stalked. The tertiary branches again can bear either solitary flowers or (shortened) branches with more than one flower (Fig. 2.12b). Side branches are usually subtended by one bract. In some species, however, two bracts instead of one are present at the base of the flowering axis. To define a character state I only considered the middle part of the inflorescence. The lowest side branches can act as if they were a main axis themselves, and also the dormant bud of the second bract can have developed. Towards the top of the inflorescence reductions are usually present. The inflorescence types are schematically represented in Figure 2.13.

In some groups of the Leguminosae the position of the inflorescence [32] is an important diagnostic feature. At first glance this seems also the case in *Spatholobus*. However, in all groups studied both axillary and terminal inflorescences were observed. In the shrubs of *Meizotropis* inflorescences are strictly terminal.

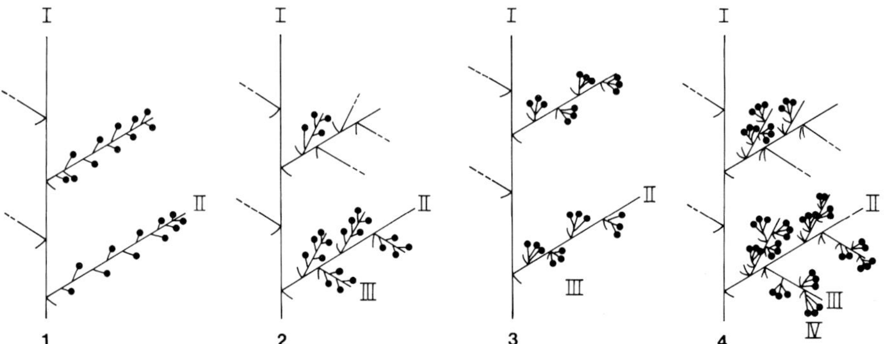

Figure 2.13. Schematic representation of the types of inflorescence. Roman numbers indicate the side branch order. The first diagrams (1, 2) with the flowers solitary on the side branches, the other (3, 4) with the flowers in brachyblasts (modified after Ridder-Numan & Wiriadinata, 1985).

In most species the indumentum of the branches of the inflorescence [33] is puberulous or pubescent. Some species have an indumentum with longer hairs, others (*S. maingayi* and in part *S. macropterus*) have more or less glabrous branches.

Species may have more than one order of side branches of the inflorescence (Fig. 2.13). The number of orders of side branches is indicated in feature [34]. The brachyblast, which can be considered a shortened side branch, represents an extra level of side branches, and is coded as 'side branch' (and not as 'flower'). This character does not indicate the presence of brachyblasts; this latter feature is coded in character 35. *Butea* and *Meizotropis* have one order of side branches. In *Butea* the side branches are shortened, but not always as short as a brachyblast. In *Meizotropis* the flowers are arranged in brachyblasts. In the species of *Spatholobus* with only one order of side branches the flowers are placed singly on branches which are never shortened (*S. acuminatus, S. albus, S. bracteolatus, S. harmandii, S. ridleyi,* and sometimes *S. multiflorus*). In *Kunstleria* there are secondary and tertiary side branches. The tertiary branches bear solitary flowers. This is also found in *S. crassifolius, S. merguensis, S. oblongifolius, S. pottingeri, S. pulcher,* and *S. sanguineus* (Fig. 2.12b). The other species of *Spatholobus* have secondary branches bearing brachyblasts (or nearly so) (Fig. 2.12a). Of these *S. littoralis* has the brachyblasts stalked. Tertiary branching with brachyblasts is found in *S. ferrugineus* and *S.parviflorus*. Inflorescences in *S. ferrugineus* can have the brachyblasts either placed on secondary or on tertiary branches.

The presence of brachyblasts [35] is a diagnostic character. One of the reasons to resurrect the species *S. sanguineus* from synonymy in *S. gyrocarpus* was the presence of brachyblasts in the former and their absence in the latter. Brachyblasts can be considered as reduced side branches. The outgroup *Kunstleria* never has the flowers placed in brachyblasts. This is in agreement with the view of Geesink (1984), who indicated a true panicle (without brachyblasts) as the less derived condition. In one species the brachyblasts are clearly stalked (*S. littoralis*), in others the brachyblasts are sessile or shortly stalked. For the analysis only the absence or presence of the brachyblasts has been coded.

The number of bracts under the first order side (= secondary) branch [36] can be either one or two. The presence of two bracts may be an indication of the reduction of a side branching (Ridder-Numan & Wiriadinata, 1985) or may be interpreted as the two stipules of a reduced leaf. This second bract is sometimes holding a dormant bud, and it is possible that a second side branch may sprout from this bud. Usually an inflorescence axis develops from the axil of one leaf or bract. In several specimens I noticed that the main axis had the side branches developing from small leaves instead of bracts. In these cases I considered the main axis in these parts to be purely vegetative, considering the side branches as axillary inflorescences, and the rest of the inflorescence as a terminal inflorescence.

The shape of the bracts of the side branches [37] is in most species either linear or triangular. Only in a few species the bracts are broadly ovate (*S. auricomus* and *S. latibractea* in part) or elliptic (*S. hirsutus*). In the latter the bracts are even surrounding the buds, whereas in other species the bud scales are the only surrounding features.

The number of bracts supporting the tertiary branch [38] is in most cases one. In case there is no tertiairy branch the number of bracts to the flower is considered. Only in *S. crassifolius*, *S. parviflorus*, *S. pottingeri* and *S. pulcher* two bracts are present.

Flower

The differences in size and shape of the flower are large between the genera, and are possibly correlated with differences in pollination. The flowers of both *Spatholobus* and *Kunstleria* are small and not brightly coloured; they have a dark red or purple to whitish colour. The pollination mechanism is not known. The flowers of *Butea*, however, are large and orange-red or yellow. It is quite possible that these large flowers are pollinated by birds. The flowers of the genus *Meizotropis* are intermediate in size relative to the other genera. The colour is red, but nothing is known of the pollinator. It is possible, however, that birds or hawkmoths are involved (see references in Ridder-Numan, 1995; in press).

Most species have a pedicel longer than 1 mm [39]. In some species, however, the flowers are nearly sessile. This is the case in most species of *Kunstleria*, in *Spatholobus gyrocarpus*, and in most specimens of *S. sanguineus*.

The bracteoles [40] are mostly placed immediately below the calyx, but sometimes they have a position lower on the pedicel. Their position is usually on the upper half of the pedicel, but in *S. acuminatus* and *S. latibractea* the bracteoles are inserted on the lower half of the pedicel. In *Kunstleria*, and usually in *Butea* and *Meizotropis* as well, the bracteoles are placed immediately below the calyx.

The indumentum of the calyx [41] is in nearly all species pubescent or sericeous (rather short-hairy, sometimes more adpressed). A few species of *Spatholobus* have a nearly glabrous calyx. A hirsute indumentum (longer hairs) is found in the species *S. pulcher* and *Meizotropis pellita*, which are more hairy over the whole plant than the other species.

The tip of the vexillary lobe [42], which comprises the two fused dorsal lobes of the five-lobed calyx, may be bidentate (Fig. 2.16e) or entire. In the latter case the tip can be rounded (or more acute) or reduced to truncate. Species with a truncate vexillary

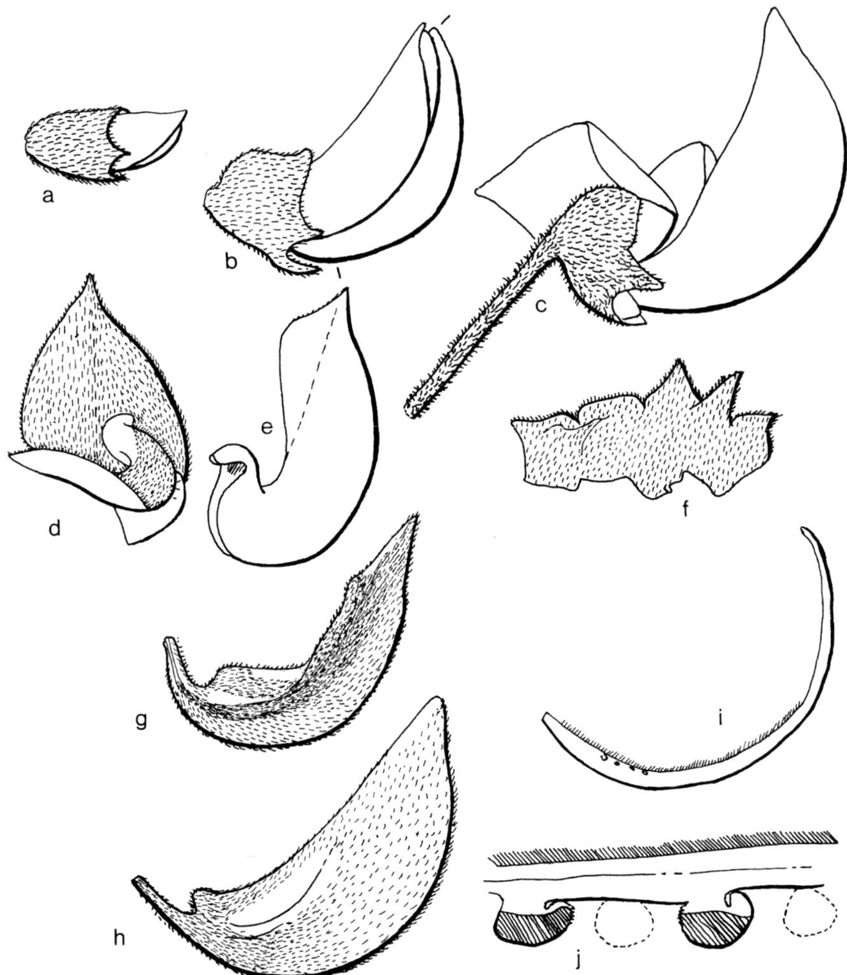

Figure 2.14. *Butea monosperma*. a–c. Flower in different stages of development; d–e. standard; f. calyx; g. wing; h. keel; i. pistil; j. ovules on one side of the ovary.

lobe are *S. apoensis, S. latistipulus, S. persicinus, S. ridleyi,* and sometimes *B. superba.* Some species are polymorphic, showing both bidentate or entire vexillary lobes.

The length of the vexillary lobe of the calyx [43] is another floral character varying between the species. In most species the vexillary lobe is about half the length of the calyx tube. In a number of species the vexillary lobe is much shorter (*S. albus, S. apoensis, S. auritus, S. hirsutus, S. latistipulus, S. littoralis, S. maingayi, S. persicinus,* and *Meizotropis buteiformis*). However, other species of *Spatholobus,* i.e. *S. acuminatus, S. bracteolatus, S. crassifolius, S. merguensis, S. multiflorus, S. parviflorus, S. purpureus* and often *S. pulcher,* have a conspicuously longer vexillary lobe. *Spatholobus pulcher* and *Butea superba* are polymorphic for this character.

The shape of the calyx lobes [44] other than the vexillary one varies between triangular and more or less rhomboid, both with acute apices, to a more or less triangular calyx lobe with a rounded apex. The short calyx lobes are often truncate or rounded.

The surface of the standard of both *Butea* and *Meizotropis* is nearly completely covered with an indumentum. Indumentum on the standard [45] is not found in *Kunstleria* or *Spatholobus*. Only one species of the outgroup, *K. kingii*, has a few hairs on the standard. I considered this last feature to be an autapomorphy for this specific species, and therefore it has been coded as absent for the genus itself.

The apex of the standard [46] is emarginate (Fig. 2.16f) in all species of *Spatholobus*, except *S. parviflorus*. The other genera have standards with an acute apex (Fig. 2.14, 2.15).

Usually the standard has no auricles or other appendages. In *Meizotropis* the standard has auricles at the lower margin of the blade, which are sometimes inflexed (Fig. 2.15d). Occasionally, the standard has auricles [47] in *S. parviflorus* and *S. ridleyi*.

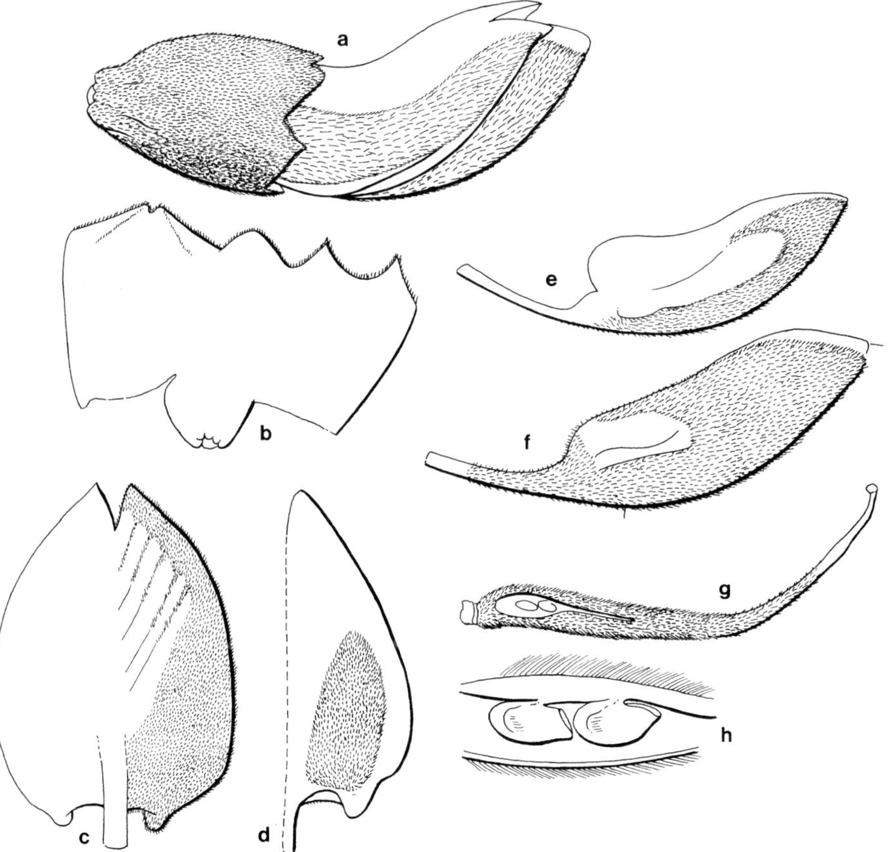

Figure 2.15. *Meizotropis buteiformis*. a. Flower; b. calyx; c. outside of the standard; d. inside of the standard; note the inflexed auricles; e. wing, f. keel; g. pistil; h. ovules in the ovary.

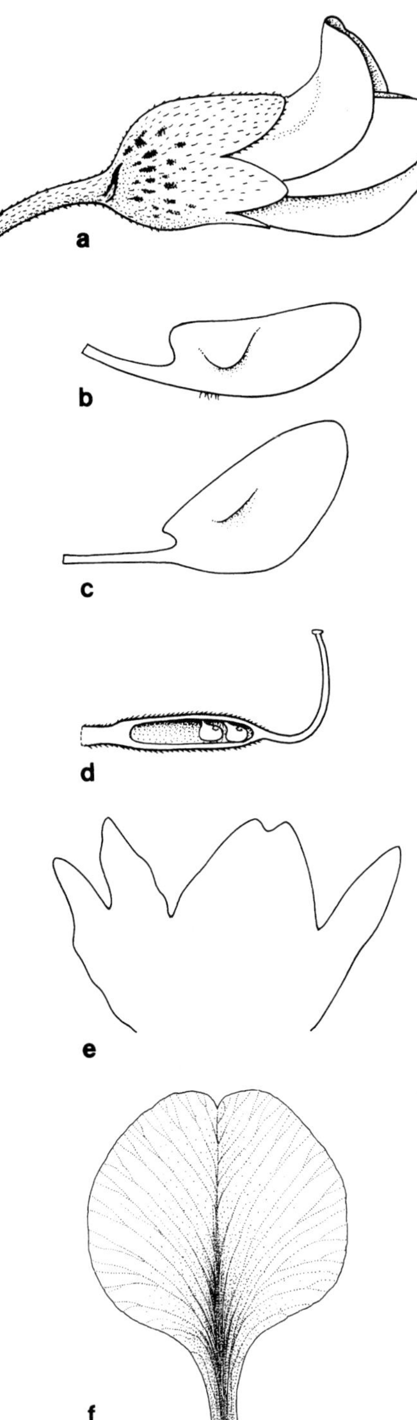

The base of the blade of the standard [48] may taper in two different ways to the claw: decurrent or truncate. In *Kunstleria* and *Meizotropis* the base is never decurrent into the claw (Fig. 2.15, 2.17), but in *Butea* and half of the species of *Spatholobus* a decurrent base is normal (Fig. 2.14d & e, 2.16f). Young petals, however, do not have a full-grown claw and are still truncate.

The shape of the blades of wing and keel petals depends on the presence of auricles on ventral and dorsal margins near the claw of the petals. The auricles of the species of *Spatholobus* are usually rather small (Fig. 2.16, 2.18), in *Kunstleria* they may be very conspicuous (Fig. 2.17c).

The dorsal auricle on the wing [49] is always present in *Butea*, *Meizotropis* and *Kunstleria*. In some species of *Kunstleria* the dorsal auricle on the wing petal may even be curled towards the claw (Fig. 2.17c). Not all species of *Spatholobus*, however, have a dorsal auricle. In *Kunstleria* and many species of *Spatholobus* the keel also has a dorsal auricle [56]. *Butea* and *Meizotropis* (Fig. 2.15) do not have a dorsal auricle on the keel petals, except for a minute one in *B. monosperma*.

A ventral auricle on the wing [50], however, is much rarer in this group. There is no ventral auricle in *Butea*,

Figure 2.16. *Spatholobus purpureus*. a. Flower; b. wing; c. keel; d. pistil showing the apical position of the ovules; e. calyx with bidentate vexillary lobe. — *S. latistipulus*. f. Standard with an emarginate apex and decurrent base.

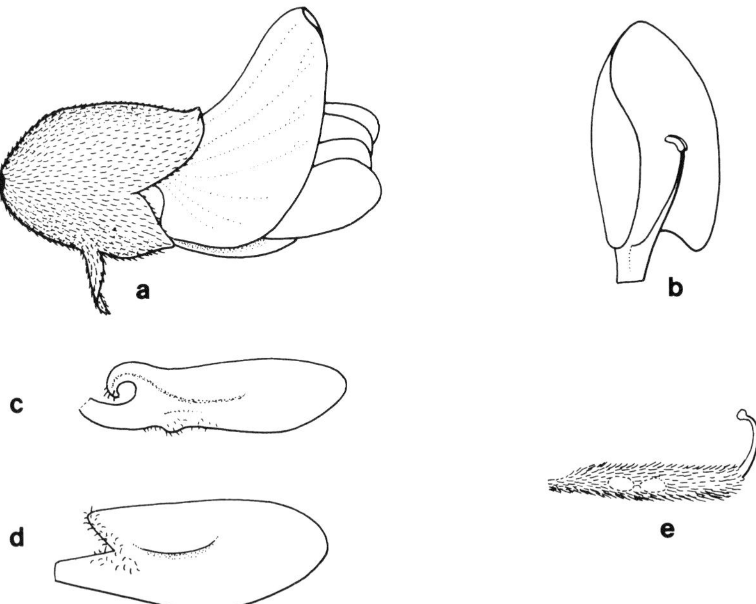

Figure 2.17. *Kunstleria sarawakensis*. a. Flower; b. standard with connate vexillary filament; c. wing with long, curled dorsal auricle; d. keel; e. pistil.

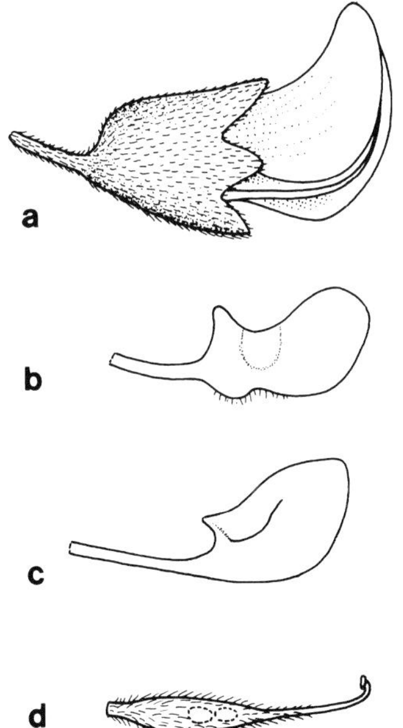

Figure 2.18. *Spatholobus ferrugineus*. a. Flower; b. wing; c. keel; d. pistil.

Meizotropis and *Kunstleria*, and it is present only in a few species of *Spatholobus*. These species, *S. albus*, *S. ferrugineus* (Fig. 2.18c), *S. multiflorus*, *S. pulcher*, and *S. viridis*, have both dorsal and ventral auricles on the wing blade. Only the last species, *S. viridis*, is sometimes without a dorsal auricle. On the keel ventral auricles are absent [57].

Lateral pockets [51] which connect the wing with the keel petals are not always present in all species of *Spatholobus*. The lateral pockets on the wing (Fig. 2.16), and on the keel [58] are always present in the other genera.

The indumentum on the petals is characteristic for several taxa. There is no indumentum on the inside of the wing petal [53] and of the keel petal [60] in *Kunstleria* and *Spatholobus*. The species of *Kunstleria* (Fig. 2.17c) usually have some pubescence on the outside of the blade of the wing petals [52].

In *Spatholobus* only a few species, i.e. *S. hirsutus*, *S. latibractea* , and *S. macropterus*, have some indumentum on the outside of the wings. In most species of *Kunstleria* the outside of the keel petal [59] has some indumentum, although this is usually restricted to the part of the blade below the auricle. In *S. latibractea* the outside of the keel petal sometimes has an indumentum. *Butea* and *Meizotropis* are rather hairy on both the inside and outside of the petals (Fig. 2.14, 2.15). *Meizotropis pellita* is glabrous on the inside of the blade of the keel and wing petals, and the inside of the keel is glabrous in *Butea monosperma*.

Cilia are present on the dorsal auricle or margin of the wing [54] and the keel [61] and on the ventral auricle or margin of the wing [55] and keel [62] in both *Kunstleria* and *Butea*, although they are much less dense in *Kunstleria* than in *Butea* (Fig. 2.14, 2.17). The two species of *Meizotropis*, however, are glabrous on the dorsal margins of both keel and wing petals (Fig. 2.15). In *Spatholobus*, only *S. latibractea*, *S. macropterus* and *S. pottingeri* have a few cilia on the dorsal and ventral margin of the wing petals. Occassionally, there are a few cilia on the ventral auricle of the wing petals in *S. ferrugineus* (Fig. 2.18b). The keel petals are glabrous in all species of *Spatholobus* except *S. latibractea*, which has a few cilia on the dorsal auricle of the keel.

The keel petals are always connate [63] to a greater or lesser extent along the lower margin to the top. In addition to being connate, the apical part of the keel petals of *Kunstleria* may slightly overlap as well. They are never free from each other.

The vexillary filament [64] is free from the other filaments, which are connate. In *Kunstleria*, however, the vexillary filament is connate at the base with the claw of the standard (Fig. 2.17b). This may be considered an autapomorphy for the outgroup; it does not occur in other genera in the Millettieae.

The surface along which the filaments are connate [65] is in most species of *Spatholobus* and *Kunstleria*, and in the genus *Meizotropis* more than half to 3/4 of the total length of the filaments. Filaments which are connate along more than 3/4 of the total filament length are present in *Butea* and in *S. hirsutus*, *S. macropterus*, *S. purpureus* and *S. ridleyi*. The other species of *Spatholobus* have their filaments connate over a much shorter length, less than half of the total filament length.

Usually there is no indumentum on the filament sheath [66], except in *Butea* and *Meizotropis*. In *Spatholobus* only one species, *S. acuminatus*, sometimes has a few hairs on the filaments.

The size of the anthers [67] is a diagnostic feature in *Butea* and *Spatholobus*. The anthers of *Meizotropis* and *B. superba* are between 1 and 2 mm long, while those of *B. monosperma* are longer than 2 mm. The anthers in the much smaller flowers of the genera *Kunstleria* and *Spatholobus*, however, are less than one millimetre in length (in most species the anther size is even less than half a millimetre). The size of the anthers was taken from mature anthers that were still carrying pollen, and in the case of alternately larger and smaller anthers the size of the larger anthers was used.

The filaments of the stamens are alternately longer and shorter. The anthers [68] are all equal in species of *Butea* and *Meizotropis* and in most species of *Spatholobus*. In *Kunstleria* the shorter filaments bear smaller anthers than the longer ones. These smaller anthers carry normal pollen grains. In some cases these pollen grains did not look

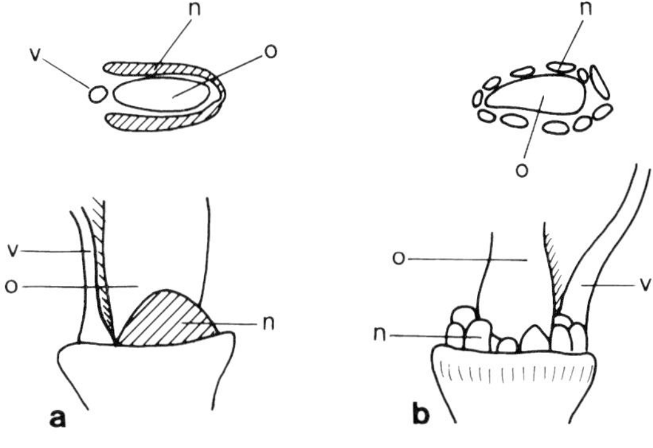

Figure 2.19. a. Ring-shaped nectary of *Spatholobus suberectus*; b. nectary consisting of 10 separate lobes (*S. ferrugineus*). n = nectary; o = ovary; v = vexillary filament.

maturewhen examined by scanning electron microscopy, but I did not test them to see whether they contained live material (Ridder-Numan, in press). Some species of *Spatholobus* – *S. auricomus, S. bracteolatus, S. ferrugineus, S. pottingeri* – also have smaller anthers, but here they are much smaller and did not contain pollen grains. This state is called sterile.

Nectary glands [69] are usually not very conspicuous. In most cases in *Spatholobus* there were ten small lobes inside the filament sheath, one at the base of each filament (Fig. 2.19b). In *S. suberectus* the nectariferous tissue is connate in a ring-like structure, and does not consist of separate lobes (Fig. 2.19a). *Spatholobus pulcher* and *S. purpureus* only have five lobes, which are situated at the base of the larger filaments. The other genera as well as some species of *Spatholobus* have no lobe-like structures, but the bases of the filaments are thickened.

The nectariferous tissue of the disk may have a short, sparse indumentum [70]. Most species, however, are glabrous.

The indumentum on the ovary [71] is pubescent or sericeous in most species, or, when denser, the appearance is more woolly. *Spatholobus latibractea* can have both a pubescent and a very densely pubescent ovary. In *S. auricomus, S. gyrocarpus, S. sanguineus* and *S. suberectus* the indumentum is sparse.

The indumentum on the ovary is continued on the style [72]. In *Butea* and some species of *Kunstleria* and *Spatholobus*, the indumentum does not reach further than the base of the style, with the rest of the style remaining glabrous. The species of *Meizotropis* and some other species of *Spatholobus* have an indumentum reaching up to at least 3/4 of the total length of the style. In the other species of *Spatholobus* and *Kunstleria* the indumentum extends halfway up the style.

The number of ovules [73] is usually two in *Kunstleria, Meizotropis*, and *Spatholobus*. The ovary in *S. pulcher*, however, may contain two to four ovules. In *Butea* there are usually more than four ovules.

The ovules are placed in the middle of the ovary or are more basal [74]. In *Butea* and *Meizotropis* they are found centrally in the ovary, while in *Kunstleria* the ovules are situated in the basal part of the ovary (Fig. 2.17e). In *Spatholobus* ovules may also be placed – apart from a median or basal situation – apically in the ovary (Fig. 2.18d). The ovary usually has a small stipe [75] in *Butea, Kunstleria* and about half the species of *Spatholobus* (Fig. 2.16, 2.17). In *Meizotropis* the ovary lacks a stipe.

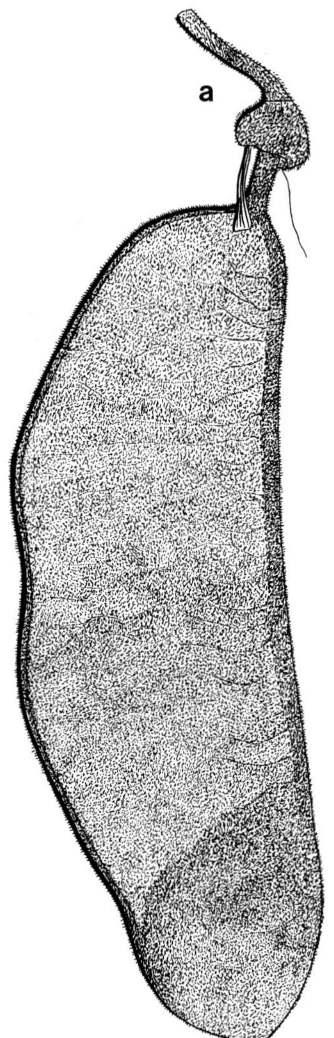

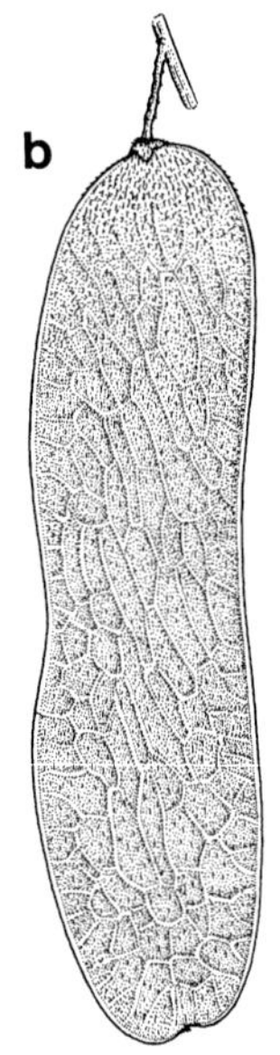

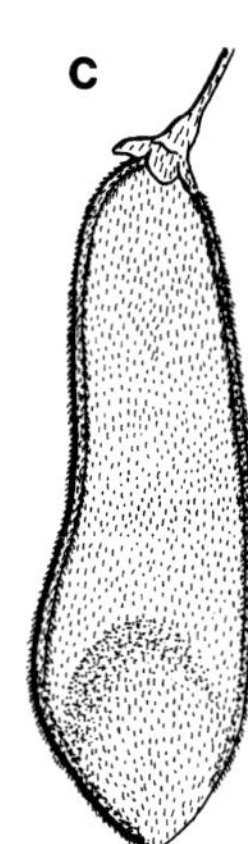

Figure 2.20. Pod of a. *Butea*, b. *Kunstleria*, c. *Meizo-tropis*.

The stigma [76] is small and may be flat, capitate or more elongated. No appendages were seen. The species of *Butea*, *Meizotropis* and *Spatholobus bracteolatus* and sometimes *S. macropterus* show an elongated stigma. The species of *Kunstleria* and about half of the species of *Spatholobus* have a capitate stigma. The other species of *Spatholobus* have a minute flat stigma.

Pod

The non-dehiscent pod of *Butea*, *Meizotropis*, and *Spatholobus* is samaroid with one apical seed (Fig. 2.20a & c; 2.21). The wing is the expanded part of the base of the ovary and may be as wide as or wider than the seed-bearing top. The wing is rather thin and the sutures on both sides are clearly visible. Only one of the ovules develops into a seed, which is connected to the dorsal suture. The style remnant is visible at the tip, sometimes pointing downwards due to the upper suture being slightly curved at the top. The pod of *Kunstleria* is strap-like and bears two very flat seeds in the middle (Fig. 2.20b).

The pod can have a stipe [77] which extends beyond the calyx. In some species there is a correlation between the presence of an ovary stipe with that of the stipe of the

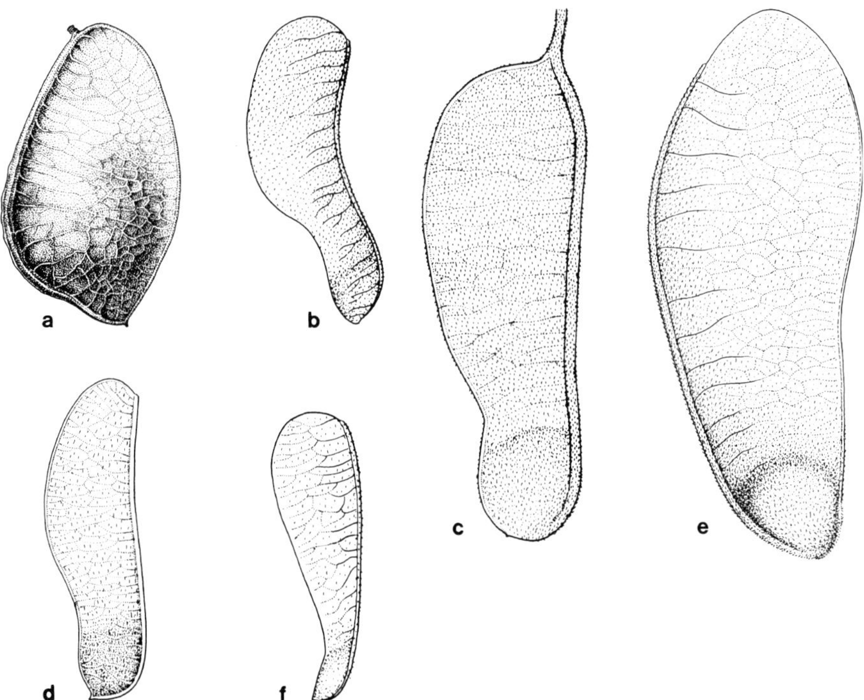

Figure 2.21. Pods of *Spatholobus*: a. *S. latistipulus*; b. *S. albus*; c. *S. parviflorus*; d. *S. oblongifolius*; e. *S. hirsutus*; f. *S. ferrugineus*. After Ridder-Numan & Wiriadinata (1985; b–f) and Ridder-Numan (1992; a).

pod, but in others there is no such correlation. The stipe of the pod does not always elongate very much when the pod expands and thus does not always exceed the calyx. The calyx is more or less persistent, and even other floral remnants may be found at the base of the pod. However, in *Butea* (Fig. 2.20a) and some species of *Spatholobus*, *S. dubius*, *S. harmandii*, *S. parviflorus* and *S. suberectus*, a conspicuous stipe is present (Fig. 2.21c). The other species of *Spatholobus*, and the genera *Kunstleria* and *Meizotropis*, have more or less sessile pods (Fig. 2.20b & c; 2.21a, b, d–f). *Spatholobus hirsutus* and *S. macropterus* show both features.

The apex of the pod [78] indicates the way the style remnant is connected to the seed-bearing top. The style can be pointing downwards or forward, or sometimes is pointing upwards. In the latter case the dorsal suture is curved downwards more than normally.

The shape of the wing [79] is one of the features of the pod. In *Spatholobus maingayi* the pod may be extremely curved with a narrow wing or straight with a wider curved wing. Usually the wing is wider than the tip and the dorsal suture is straight. Sometimes the wing is not wider than the seed-bearing top with the pod: *S. acuminatus*, *S. latistipulus* (Fig. 2.21a), *S. littoralis*, *S. maingayi*, *S. oblongifolius* (Fig. 2.21d) and *Kunstleria* (Fig. 2.20b).

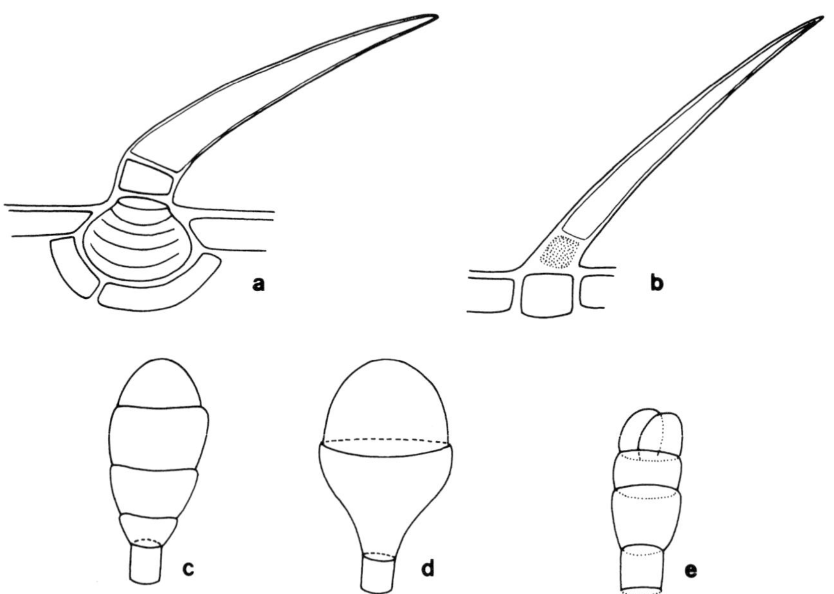

Figure 2.22. a. Non-glandular uniseriate hair of *Spatholobus*; b. non-glandular uniseriate hair of *Butea*; c. glandular hair of *Spatholobus*; d. glandular hair of *S. parviflorus*; e. glandular hair of *Butea*.

Leaf anatomy

The leaves are dorsiventral with one to three palisade layers on the dorsal side, a central layer and one to three layers of spongy parenchyma. Both glandular and non-glandular hairs are present.

The non-glandular hairs are uniseriate, with one long straight or slightly curved top cell and a few short basal cells (Fig. 2.22a & b). The hair base is often surrounded by epidermal cells in a rosette (Plate 2.1a). The base of the hair [81] is either sunken in the palisade layer (Fig. 2.22a, Plate 2.1b) or is situated in the epidermal layer and not sunken into the lower layers (Fig. 2.22b). In *Spatholobus* both types of hair bases are present, often in the same species. Of *Spatholobus* only *S. parviflorus* and *S. pulcher* do not have hairs with a base sunken in the palisade layer. In 13 species the character is polymorphic. *Butea*, *Kunstleria* and *Meizotropis* never have their hair bases sunken in the palisade layer.

The presence of a bulbous septate hair base [82] seems to be dependent on the presence of hairs with a base sunken below the epidermal layer (Fig. 2.22a, Plate 2.1b). Species that never have the hair base sunken in the palisade layer also never show a bulbous septate hair base. However, *Kunstleria*, which has no basal cells of the hairs sunken in the palisade layer, has both hairs with or without a bulbous septate base. Because this character turned out to be too much associated with the former character [81] and because of the relatively large number of species I decided to use only presence or absence of a bulbous septate hair base.

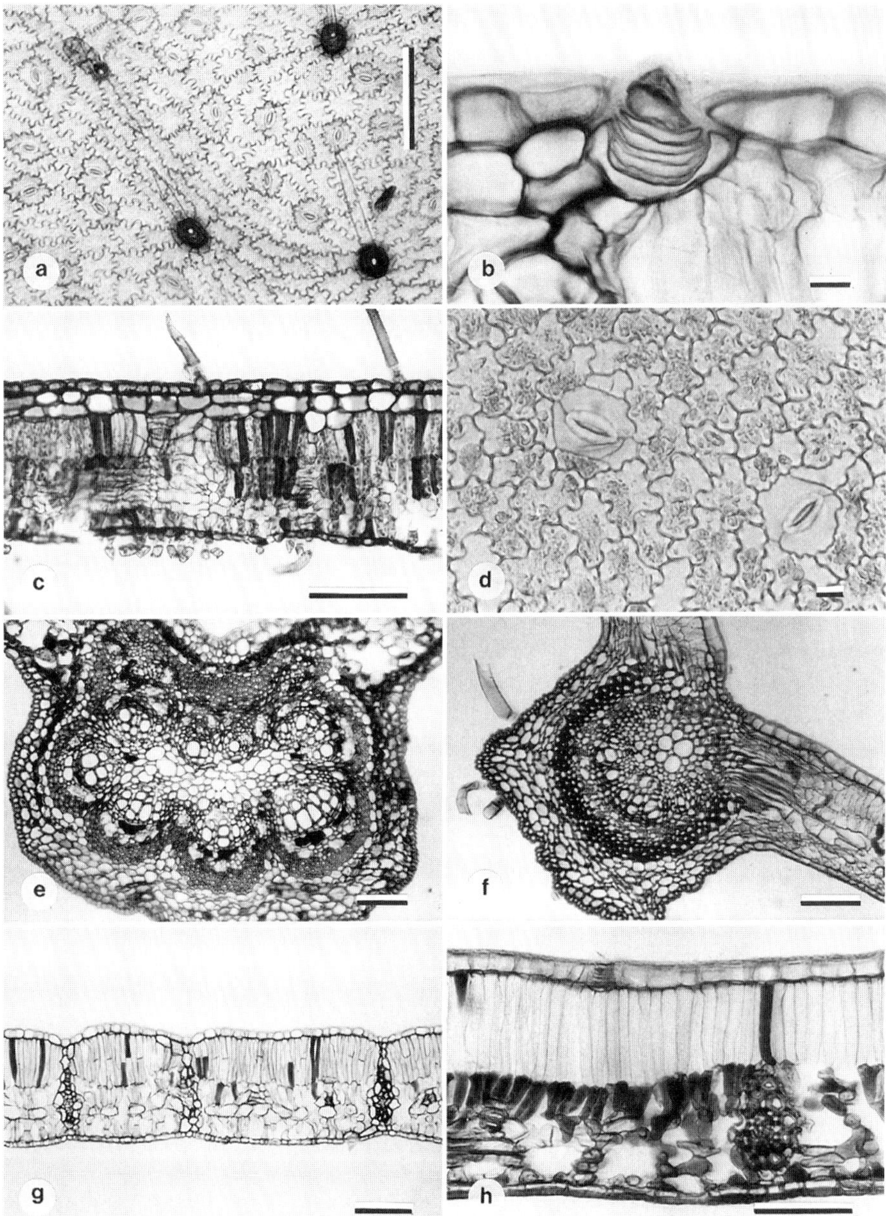

Plate 2.1. – a. Hairbase in surface view with epidermal cells in a rosette; note the undulating anti-clinal cell walls (*Spatholobus acuminatus*). – b. Bulbous septate hair base in cross section of a leaf of *S. littoralis*. – c. Cross section of a leaf of *S. parviflorus*; note the hypodermis. – d. Adaxial surface of a leaflet of *S. macropterus* showing stomata. – e. Midrib of a leaflet of *S. macropterus* showing the ring-like shape of the vascular bundle; note the dark substance, presumed to be tannin. – f. Midrib of *Kunstleria ridleyi* showing an arch-shaped vascular bundle. – g. Transcurrent veins in a cross section of a leaf of *Spatholobus pulcher*. – h. Continuous palisade layer in a cross section of a leaf of *Kunstleria kingii*. — Scale bars of a, c, f–h = 0.1 mm; of b, d, e = 0.01 mm.

Glandular hairs [83] are present in all species of *Spatholobus*, *Butea*, and *Meizotropis*, except in *S. crassifolius*. The glandular hairs are club-shaped and consist of a stalk and a head (Fig. 2.22c–e). The glands of *Kunstleria* have a longer stalk and head than those of *Butea*, *Meizotropis* and *Spatholobus*; *S. parviflorus* has a slightly aberrant glandular hair with the head divided into two parts; the lower half is holding – like an egg cup – the thinner walled upper half (Fig. 2.22d).

On both leaf surfaces the anticlinal walls of the epidermis show a specific shape. The walls can be straight, curved, or more or less strongly undulated (Plate 2.1c). The stronger undulations give the cells the impression of pieces in a jigsaw puzzle, and the strongest ones are those with undulations in the shape of an omega (Ω), where the tops of the undulation are (nearly) touching each other. The walls at the abaxial side [84] are usually more curved than the adaxial ones. For the analysis I used only the anticlinal cell walls at the abaxial side. This character may be influenced by the amount of light, and sun and shade leaves of the same plants may differ (shade leaves usually have more undulated cell walls). This is probably the reason why this character is polymorphic for five species of *Spatholobus*. A short discussion can be found in Metcalfe & Chalk (1979). A feature that makes it difficult to see the shape of the anticlinal wall clearly is the difference in undulation between inner and outer wall of the same cell.

In most species a real hypodermis [85] is absent. In some species there is sometimes a part near the larger veins where there is an extra layer between epidermis and palisade layer. *Spatholobus ferrugineus* always shows this feature near the larger veins; *S. parviflorus*, however, has a continuous layer of one or two translucent cells under the epidermis (Plate 2.1c). The presence of a continuous hypodermis in *S. parviflorus* seems to be an autapomorphy.

The stomata of all species are paracytic and usually not sunken. Adaxial stomata [86] are absent in *Butea*, *Meizotropis*, and *Kunstleria*. In *Spatholobus* the presence of adaxial stomata is not uncommon. They are absent in *S. crassifolius*, *S. dubius*, *S. gyrocarpus*, *S. multiflorus*, *S. parviflorus*, *S. pottingeri*, *S. pulcher* , *S. purpureus* and *S. suberectus* In some other species they are not always present. Adaxial stomata are very frequent in *S. hirsutus*, *S. latibractea*, *S. oblongifolius*, and *S. viridis*, and in some specimens of *S. albus*, *S. apoensis*, *S. latistipulus* and *S. macropterus* (Plate 2.1d).

In cross section the vascular system in the midrib [87] is usually a closed ring (Plate 2.1e). Only in *Kunstleria* the open arch-shaped type is found (Plate 2.1f). In some species of *Spatholobus* this feature is also present, however not in all specimens: i.e., *S. ferrugineus*, *S. gyrocarpus*, *S. harmandii*, and *S. sanguineus*.

Transcurrent veins [88] are present in several species. The veins can be transcurrent to the upper and lower epidermis by bundle sheath extensions or by the bundle sheath itself (Plate 2.1g). When the palisade layer is continuous above the larger veins, transcurrent veins were scored absent (Plate 2.1h).

Sometimes a cell layer with dark reddish brown contents, presumed to be tannins, occurs between phloem and xylem [89] in *Spatholobus* (*S. auricomus*, *S. dubius*, *S. multiflorus*, and *S. oblongifolius*) (Plate 2.1f). More often this layer was absent, or situated in the xylem itself. In *Butea*, *Meizotropis*, and *Kunstleria* the layer was always present between phloem and xylem.

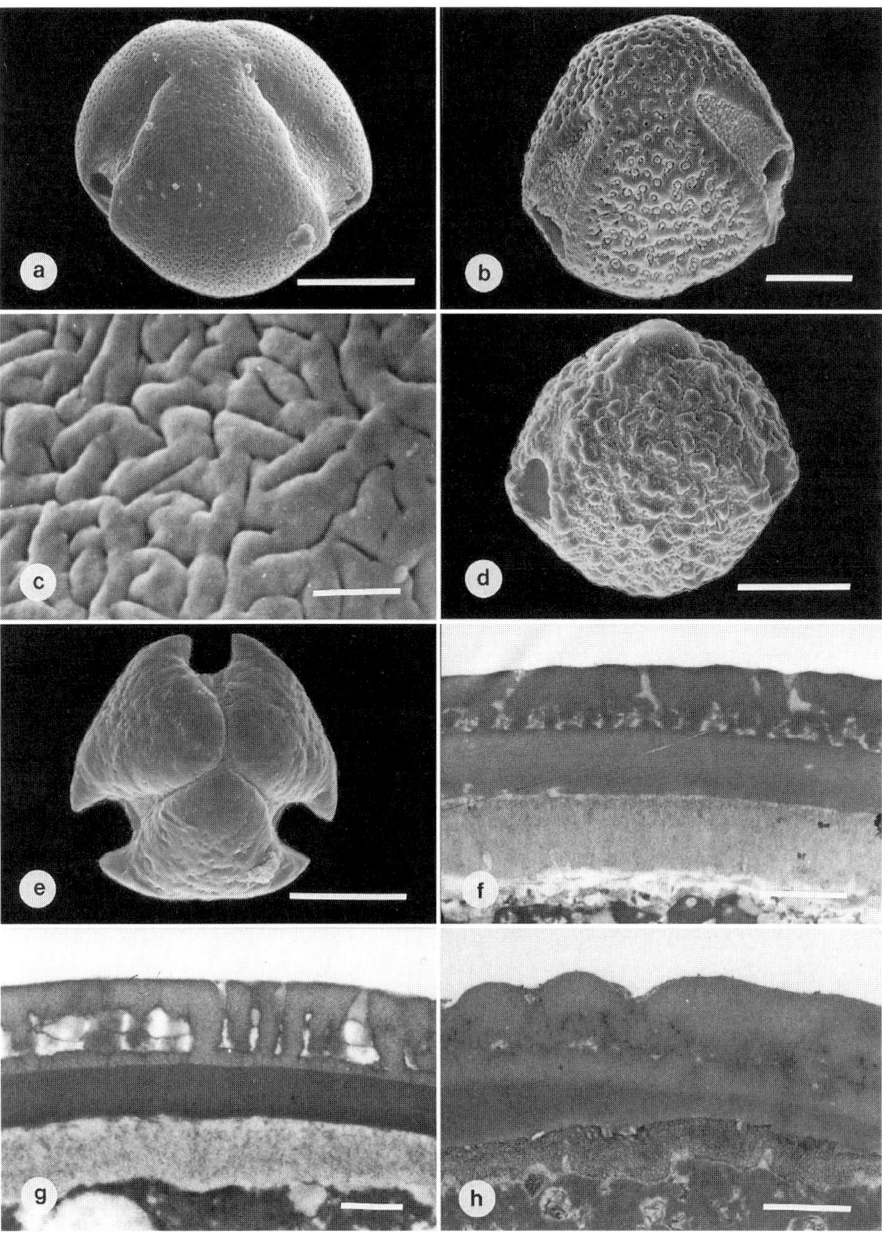

Plate 2.2. – a. Ornamentation psilate-perforate (*Spatholobus ferrugineus*); note the acute colpus ends. – b. Ornamentation microreticulate (*Butea monosperma*); note the obtuse colpus ends. – c. Rugulate ornamentation (*Kunstleria ridleyi*). – d. Verrucate ornamentation (*Meizotropis pellita*). – e. Fused colpi (*Spatholobus oblongifolius*). – f. TEM section of a pollen grain of *S. parviflorus* showing the very thin foot layer. – g. TEM section of a pollen grain of *S. ferrugineus* showing the relatively thick foot layer and the well-spaced columellae. – h. TEM section of pollen of *S. littoralis*. Note the thick foot layer and the granule-like infratectum. — Scale bars of a, b, d, e = 10 µm; of c, f–h = 1 µm.

Dark substances were also present in the bundle sheath cells [90] of some taxa. This was never the case in *Kunstleria* and *Meizotropis*, but always in *Butea* and some species of *Spatholobus*.

Pollen

The pollen grains are (sub)spheroidal and usually tricolporate, but (para)syncolporate pollen is found in several species of *Spatholobus*. For the analysis I used characters of the ornamentation, the ectoaperture morphology and the exine structure of the pollen. For a more extensive description of the pollen morphology see Ridder-Numan (in press).

The ornamentation of the pollen [91] is psilate-perforate to fossulate in most species of *Spatholobus* (Plate 2.2a). In some species the ornamentation is more open, showing elements of the infratectal layer. A more open ornamentation is present as well in *Butea* and *Meizotropis* (Plate 2.2b). Here the perforations may be a little wider, showing one or two, rarely up to seven free sexine elements (microreticulate). *Kunstleria* has a closed, usually rugulate ornamentation (Plate 2.2c). A verrucate ornamentation is present in *Meizotropis pellita*, and sometimes in *M. buteiformis* and *Spatholobus maingayi* (Plate 2.2d).

Fused colpi [92] are more or less frequent in some species of *Spatholobus*, *S. harmandii*, *S. persicinus*, *S. viridis* (Plate 2.2e). However, (para)syncolpi never occurred in all grains of one sample. The apocolpium is in this case reduced by the fused colpi, or sometimes the colpi fuse in such a way that the apocolpium nearly becomes an island. In the other three genera no (para)syncolporate grains were observed.

Colpus ends [94] may be acute (Plate 2.2a) or obtuse (Plate 2.2c). Obtuse ends are present in *Butea* and *Meizotropis*. In some species of *Spatholobus* both acute and obtuse colpus ends were found. In the one sample of *S. auritus* I noticed only obtuse ends.

The A/E-index is a measure of the apocolpium size [97], in which A is the distance between two of the three colpi and E the width of the pollen grain in equatorial view. I took this measure from pollen grains that did not have fused colpi, which have an A/E-index of zero. When fused colpi were present in a sample, the A/E-index was low in most cases, as expected. In *Spatholobus ferrugineus*, however, the colpi were never fused and the A/E-index was low. If the A/E-index is high the colpi are relatively short. In *S. gyrocarpus*, *S. sanguineus*, and especially *S. pulcher* the colpi are short and the A/E-index is high. The other three genera also have relatively short colpi. The A/E-index values are given in Figure 2.23.

Of the exine stratification the relative size of the foot layer and the structure of the infratectal layer are the two characters chosen for the analysis. Unfortunately it was not possible to make TEM sections of all species; therefore several species had to be given a '?' for these characters.

The foot layer [93] is very thin (less than 1/10 of the endexine) in *Meizotropis* and *Spatholobus parviflorus* (Plate 2.2f). A thicker foot layer is present in *Butea* and most species of *Spatholobus* (Plate 2.2g). A relatively thick endexine associated with the reduction of the foot layer may be regarded as a morphologically advanced feature in the Papilionoideae (Ferguson & Skvarla, 1981; Ferguson, 1984). A foot layer thicker

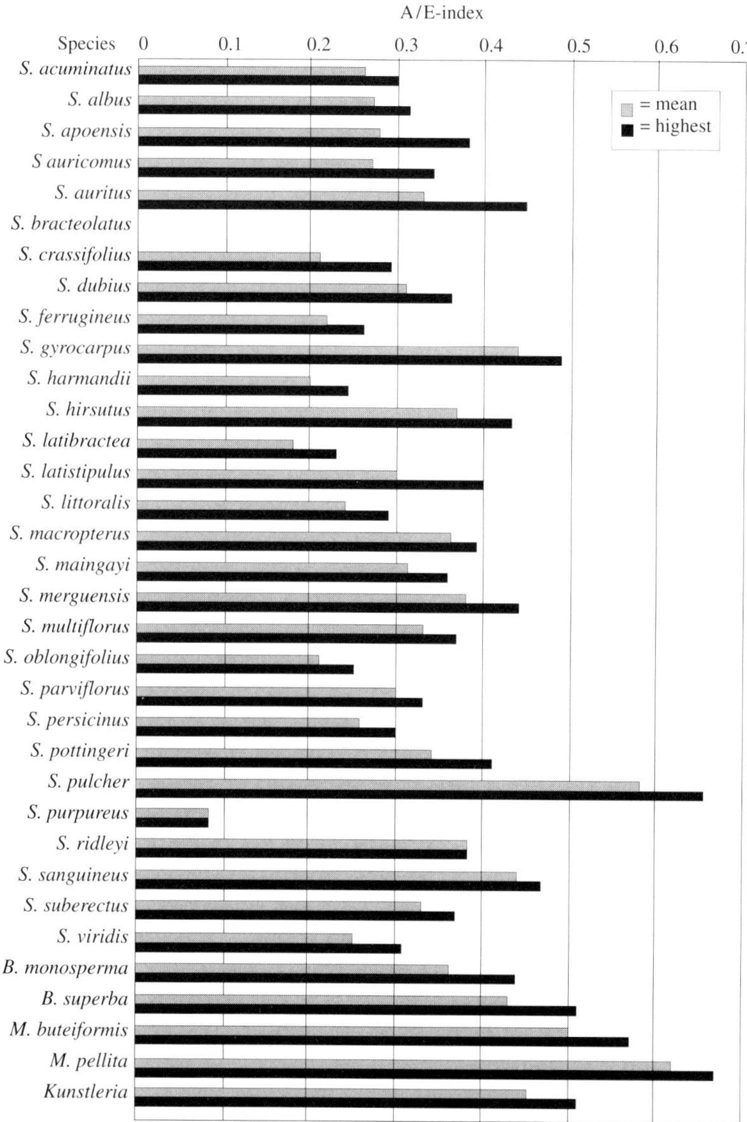

Figure 2.23. A/E-index of the pollen grains. Pollen of *Spatholobus bracteolatus* was not available.

than half the endexine, even sometimes thicker than the endexine, was found in *Kunstleria* and some other species of *Spatholobus*. The foot layer–endexine ratio of the mean values of foot layer and endexine was used in the analysis (Fig. 2.24; only species of which a TEM section was available are included).

The infratectal layer [97] is a difficult character to code, resulting in most species being polymorphic for two or three states. Especially when the infratectal layer is thin,

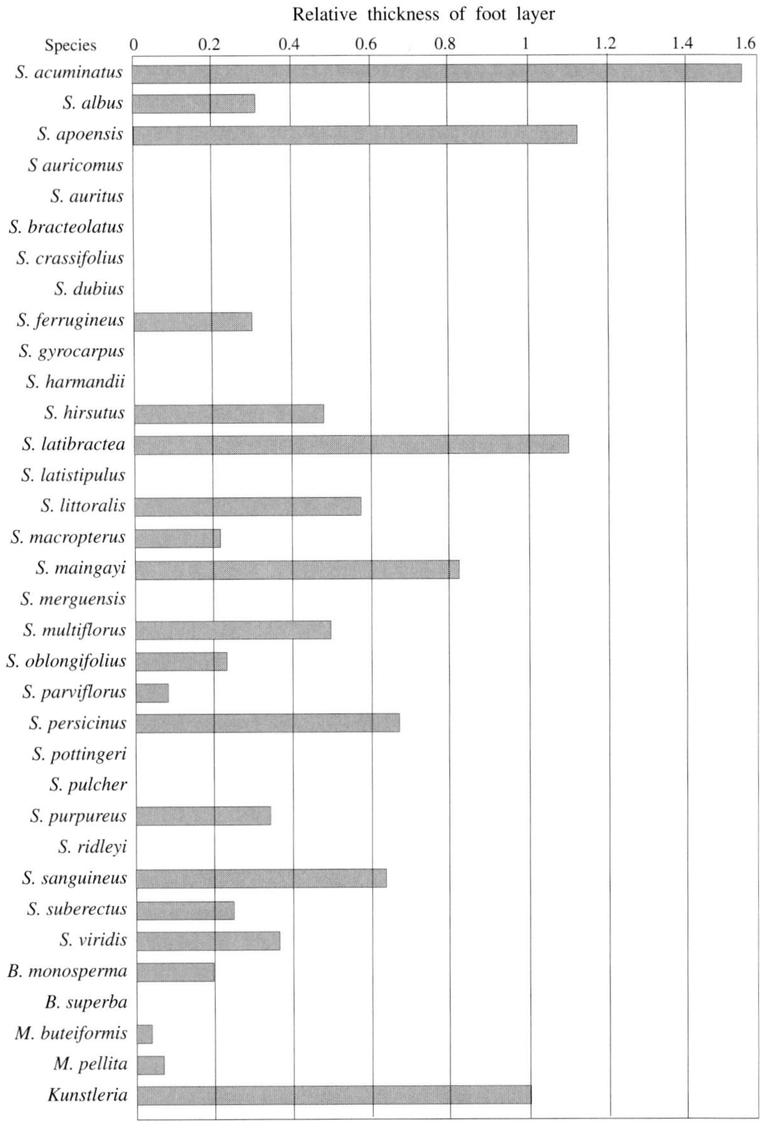

Figure 2.24. Thickness of the foot layer relative to the endexine. Values are given only for species of which TEM data are available.

which is so in most species, the columellae will be short and the difference between a granular and a columellate infratectal layer is not easy to see (Plate 2.2h).

Mesocolpial pouches [96] can be found in most species. In *Butea*, *Meizotropis*, *Spatholobus parviflorus* and *S. pulcher* this feature was present in very few grains. It is possible that in the case of a thin and flexible pollen wall, the wall may fold more easily into mesocolpial pouches.

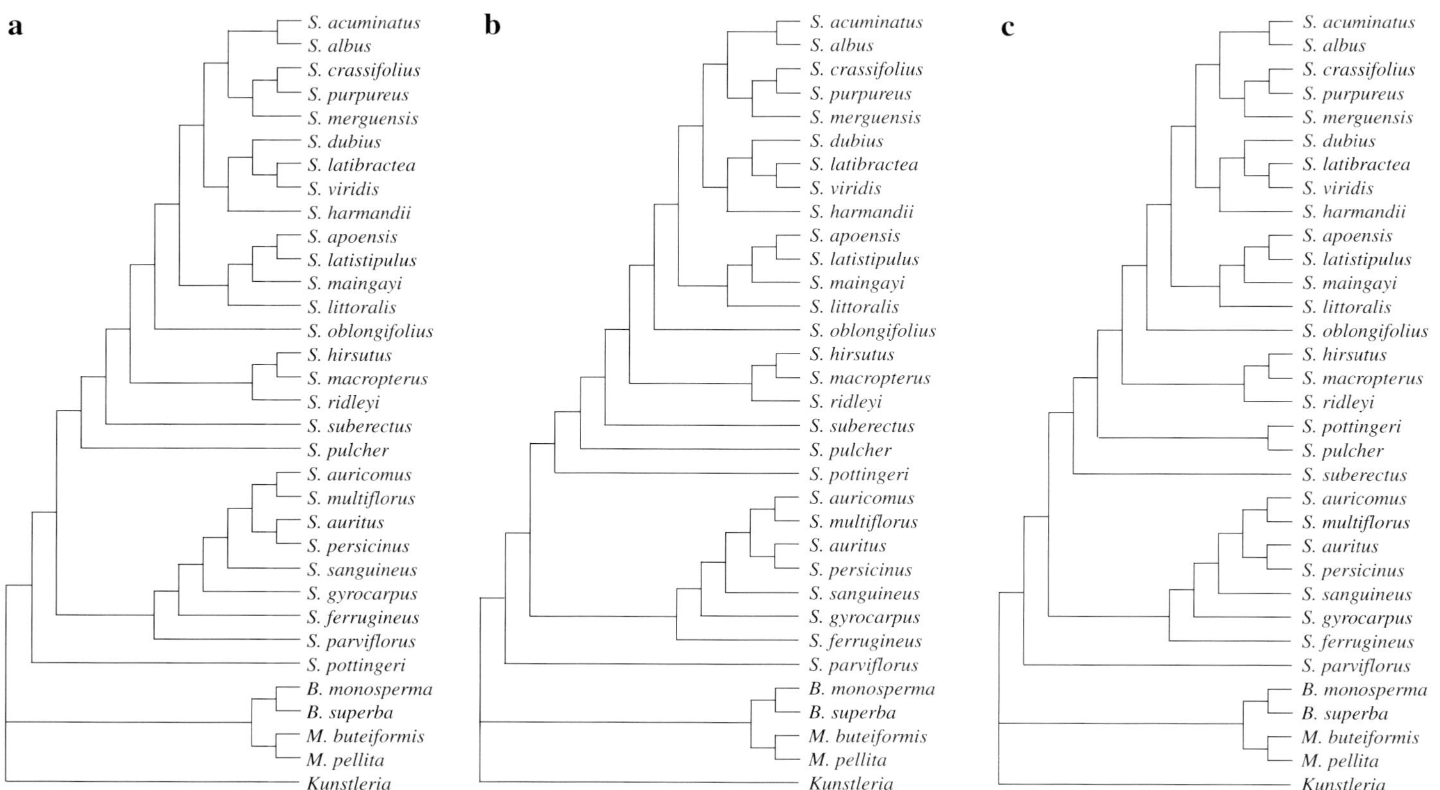

Figure 2.25. All three most parsimonious trees found by using PAUP under the heuristic search option, 'stepwise addition sequence random', tree-bisection-reconnections (TBR); all characters analysed unordered, uninformative characters ignored and characters with multiple states treated as polymorphisms; length 589, c.i. 0.46. On each of the branches the amount of change is shown. a = tree 1; b = tree 2; c = tree 3.

RESULTS

Finding the set of most parsimonious trees

Following the phases of preliminary analyses, checking and rechecking data, coding and recoding characters, a final phylogenetic analysis was performed. This analysis was carried out using PAUP under the heuristic search option with an addition sequence random, and the branch-swapping procedure TBR, using *Kunstleria* as the outgroup. One species, *Spatholobus bracteolatus*, was excluded from the analysis because of the lack of sufficient data.

The analysis resulted in three most parsimonious trees (MPT) with a length of 589 steps, and a consistency index of 0.46 (Fig. 2.25). The consistency index is, according to Sanderson and Donoghue (1989), within the expected ranges for analysis with this number of taxa involved.

An evaluation of 10,000 random trees from the set of all possible trees with the same options as above gave a frequency distribution of the tree length in number of steps as shown in Figure 2.26. The three MPT's (length = 589 steps) found were much

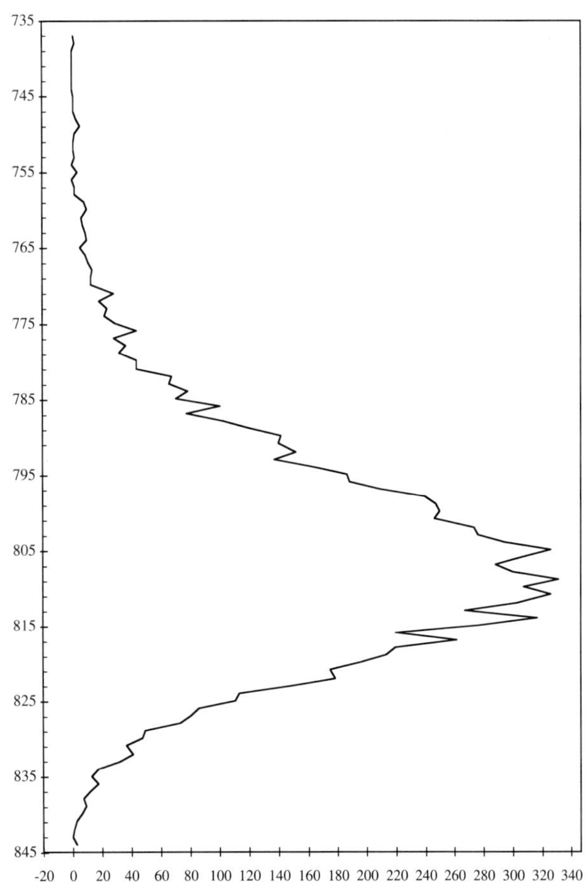

Figure 2.26. Frequency distribution of lengths of 10,000 random trees (mean = 805.6, sd = 13.7, g1 = –0.57, g2 = 0.783).

shorter than the random tree set, which had a mean value of 805.6 steps, and a standard deviation of 13.7. It is unlikely that the MPT's are randomly found trees, and it is very probable that the data set has more structure than would be expected by chance alone.

Another type of analysis is the parsimony analysis with implied weights. In the computer program PeeWee (Goloboff, 1993) the trees with the lowest number of extra steps in a character (best fit) are selected instead of the ones with the lowest number of steps (most parsimonious trees). Weighting in this way is carried out during the analysis. With this program the data were analysed using three different types of concaveness: 1, 3, and 6. However, it is difficult to know which concavity index to use [see also the discussion in Turner (1995) and Turner & Zandee (1995)]. The fit of the three trees found with concaveness 3 was 447.4, that of the two trees resulting from the analysis with concaveness 6 was 567.2, and that of the one tree found with concaveness 1 (which is equal to not weighting) was 285.8. When the statistics of MacClade where used to make it comparable to the most parsimonious trees found with PAUP these resulted in trees with length 827 and 823, and ci = 0.35. These lengths are nearly within the range of the mean value of random tree lengths that was found during evaluation of 10,000 random trees. The trees have some interesting features, but most of the clades seem to be very unnatural. The genera in the analysis turn out to be not monophyletic as one of the two species of *Butea* is placed as sister to *Spatholobus ferrugineus* halfway through the tree, and the other at the base as sister to one of the two species of *Meizotropis*. The other species of *Meizotropis* is placed as sister to *Kunstleria*. This outcome is very improbable, because in my view the monophyly of these two small genera is not in dispute.

Choice among the most parsimonious trees

The result of extensive reciprocal illumination and the running of several analyses turned out to be in certain respects rather stable. Mostly changes in the data matrix resulted in different trees. However, in most MPT's found during several trials the following more or less identical groups of species were found:

Spatholobus apoensis, S. latistipulus, S. littoralis, and *S. maingayi,*
S. acuminatus and *S. albus,*
S. crassifolius, S. merguensis (and often *S. purpureus*),
S. dubius, S. harmandii, S. latibractea, and *S. viridis,*
S. hirsutus, S. macropterus, S. ridleyi, sometimes including *S. oblongifolius* and
    S. pottingeri,
S. auricomus and *S. multiflorus,*
S. ferrugineus, S. gyrocarpus, and *S. sanguineus,*
Butea and *Meizotropis.*

Spatholobus parviflorus was always present near the base of the *Spatholobus* part of the cladogram. A few species and small clades, however, occurred in different places in the different cladograms (probably due to missing data or polymorphic states). Wandering species and small groups were *S. auritus, S. persicinus, S. pottingeri, S. pulcher* and *S. suberectus,* and the twins *S. auricomus* and *S. multiflorus.* The wanderers were usually placed in the lower half of the cladogram, and the twins have been placed either near the top of the cladogram (mostly in the *S. harmandii*-clade) or at the end of the *S. ferrugineus*-clade.

The MPT's found in the final analysis show large similarities regarding the groups mentioned (Fig. 2.25). In one of the three trees (Fig. 2.25a) *S. pottingeri* is placed in a basal position and *S. parviflorus* is inserted into the clade of *S. ferrugineus*. However, on account of similarities in leaves and pods, the position of *S. parviflorus* near the base of the cladogram next to the *Butea*-clade seems more acceptable to me than within the group of *S. ferrugineus*. Besides, the flower structure is quite different from that in the *S. ferrugineus*-group. Similarly, the basal position of *S. pottingeri* in the cladogram next to the *Butea*-clade and far from *S. suberectus*, which it resembles very much in overall characteristics, is not preferable. In the second and third cladogram (Fig. 2.25b & c) *S. parviflorus* is placed next to the *Butea*-clade in a solitary position, and *S. pottingeri* can be found in the middle of the cladogram next to *S. pulcher*, and only one node from *S. suberectus*. The other 'wanderers', the twins *S. auricomus* and *S. multiflorus*, and the species *S. auritus* and *S. persicinus*, are placed in the top of the *S. ferrugineus*-clade in all three trees. *Spatholobus auricomus* and *S. multiflorus* are species with extremely small flowers. *Spatholobus sanguineus* and *S. gyrocarpus*, both belonging to the *S. ferrugineus*-clade, have the same type of small flower. In this view one would expect these two species to be close to *S. sanguineus* and *S. gyrocarpus* as is the case in the MPT's in Figure 2.25.

There are several ways to choose between equally parsimonious trees. One of the often used methods is successive weighting. Successive weighting resulted in one tree slightly different from the set of three MPT's found as result of the initial analysis. The most parsimonious tree found by successive weighting (TBR, random 100 repetitions) is one step longer than the original set of most parsimonious trees: 590 versus 589 steps (Fig. 2.27). This tree is nearly identical to the second of the three MPT's found without weighting (Fig. 2.25b); the only difference is the position of the *S. maingayi*-clade between the *S. crassifolius*-group and the *S. acuminatus*-group. This indicates a preference for the second tree in Figure 2.25.

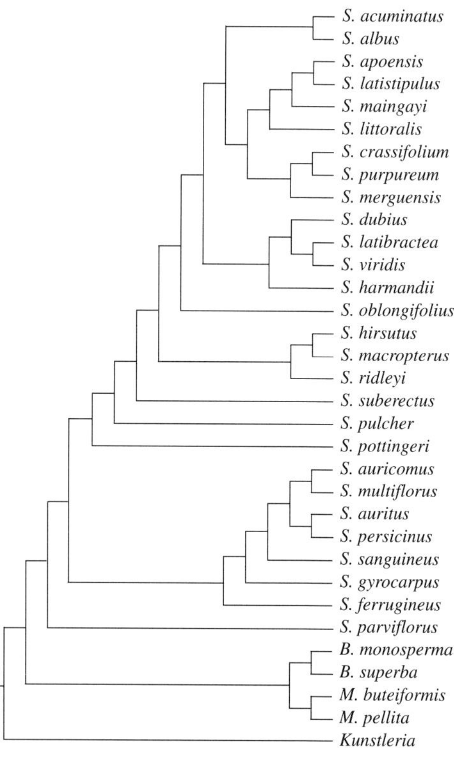

Figure 2.27. The most parsimonious tree obtained by successive weighting. Length 590, c.i. 0.46, r.i. 0.5, r.c. 0.23.

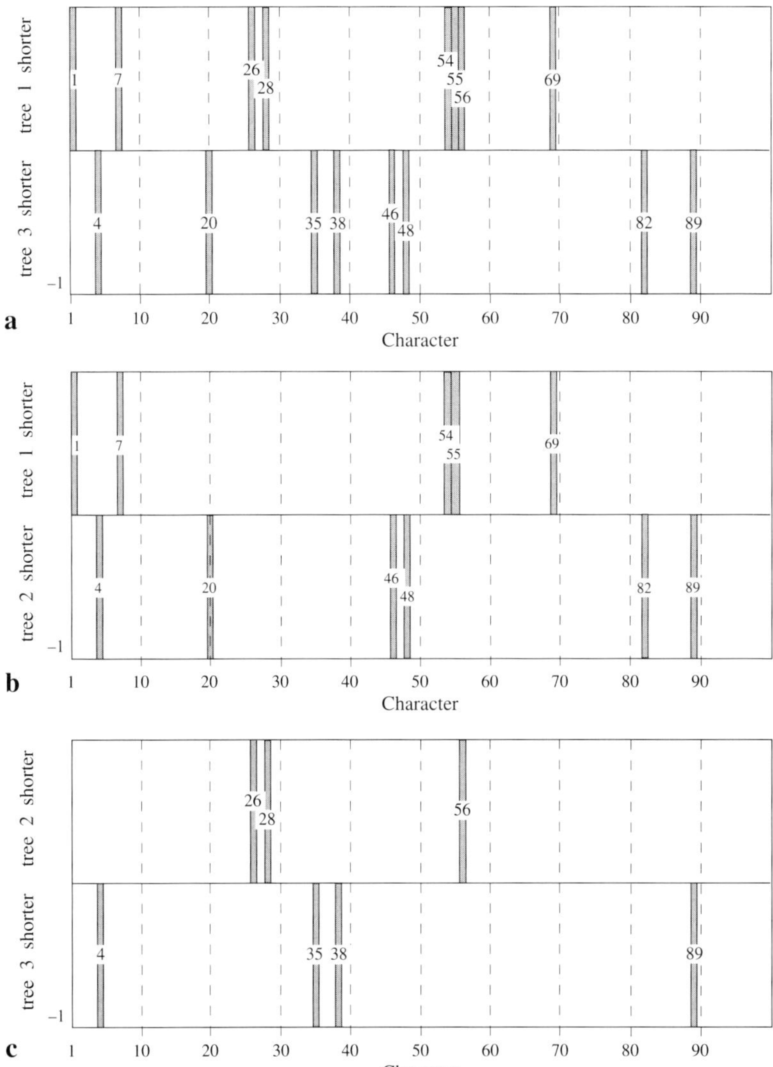

Figure 2.28. Differences between the MPT's in amount of change per character. a. Tree 1 compared with tree 3; b. tree 1 compared with tree 2; c. tree 2 compared with tree 3.

When taking into account the amount of support for the branches in the three different trees, there are no big differences (MacClade: Trace all changes and All change options: unambiguous changes only, graphics display: label by amount of change). Except for some other slight changes, tree two has more support for the first branch leading to the genus *Spatholobus*.

The difference between the three trees in amount of change per character is illustrated in Figure 2.28 (MacClade: Chart compare two trees). In this figure the differences between two trees in amount of steps per character are given. Compared with the first tree (= tree 1) the third tree (= tree 3) is one step shorter for 8 characters, in all other characters they have the same length. The characters for which tree 1 is shorter are: 1, 7, 26, 28, 54, 55, 56, 69. Tree 3 is one step shorter from tree 1 in the characters: 4, 20, 35, 38, 46, 48, 82, 89. Comparison between the second of the MPT's (= tree 2) and tree 3, shows that tree 2 is shorter for 3 characters (26, 28, 56). Tree 3 is different from tree 2 in 3 characters (4, 35, 38). The characters for which tree 2 is shorter than tree 1 are: 20, 46, 48, 82, 89. Tree 1 is shorter than tree 2 for the characters 1, 7, 54, 55, 69. Tree 2 is less different from the other two trees than the others are from each other.

Using the weighting method as described by Turner (1995), this gives for tree 1 W = 40.71, for tree 2 W = 40.63, and for tree 3 W = 40.76. This weighting method is in fact a comparison between the sum of the retention indices for all individual characters for each of the weighted trees:

$$W = \sum ri, \text{ where } ri = (Es_{i,max} - Es_i) / Es_{i,max}$$

$Es_{i,max}$ = maximum number of extra steps for character i on the tree (i.e. maximum–minimum number of steps)

Es_i = observed number of extra steps for character i (i.e. observed number–minimum number of steps)

The tree with the highest fit (ri = 1, is the best fit; ri = 0, is the worst fit) will have the highest W, because the homoplasy is concentrated in the fewest characters. In this case tree 3 has the highest value of W. The values, however, do not differ very much. This is understandable, because in the tree comparison by amount of change (Fig. 2.28), none of the trees differ in the number of characters for which they are shorter than any of the other two trees. The homoplasy is concentrated in the same number of characters. The weighting method of Turner is not very discriminating in this case. In a former analysis with a slightly different data matrix, and which resulted in three equally parsimonious trees as well, there was a difference between the three trees in the number of characters in which the homoplasy was concentrated. In this case the tree that had the homoplasy concentrated in the fewest characters showed the highest value of W. In my view the two described methods above both end up with the same result, because they are both looking at the concentration of homoplasy in the MPT's, i.e. the higher the W the better the fit according to the method of Turner, and the fewer characters showing the homoplasy (= the fewer characters shorter for one of the MPT's) the better the fit of the tree according to the comparison of two trees in MacClade. The last method is very simple when few MPT's are involved, but Turner's method is easier with larger numbers of MPT's (in CAFCA this option is available).

As there are only three equal MPT's – two of which are hardly different – it may be possible to select one of the trees on morphological arguments. *Spatholobus parviflorus* differs from the other species in *Spatholobus*, and more closely resembles *Butea*. It is in various respects similar to *Butea*, e.g., a large pod, shape and structure of the leaves, extremely long calyx lobes, and a standard apex that is acute. Although in all other aspects it certainly belongs to *Spatholobus*, a place in the cladogram as close to *Butea* as possible would be expected. In the first cladogram *S. parviflorus* is included in the

S. ferrugineus-clade, in the second and third cladogram it is placed at the base of the *Spatholobus* clade. The latter is preferable in my view. The other differences between the three MPT's concern the position of *S. pottingeri*. In the second and third cladogram this species is placed as sister to *S. pulcher*. In the first cladogram, however, *S. pottingeri* is placed basally in the *Spatholobus*-clade, while at the same time *S. parviflorus* occurs in the previously mentioned, less preferable position within the *S. ferrugineus*-clade. Based on overall similarity, but especially the general appearance of the flower (shape and size of calyx and petals), the species *S. pottingeri*, *S. pulcher* and *S. suberectus* appear to be closely related. At first sight it is not easy to tell them apart, but the differences become more apparent in detail. These considerations favour the choice for the second or third tree. Tree 2 and tree 3 only differ in the position of *S. pottingeri* with respect to *S. pulcher* and *S. suberectus*. There are as many morphological arguments to place these three species in the position in tree 2 as for the position they take in tree 3.

As a result of the considerations in this paragraph, especially the morphology and the result of weighting, tree 1 was discarded. Because there are no convincing arguments to chose between the remaining two, I accepted tree 2 as a working hypothesis (Fig. 2.25b).

Evaluation of the tree

Character support

Based on the arguments given above I accepted the second of the MPT's in Figure 2.25 as a working hypothesis. Several branches of smaller and larger parts of this tree (Fig. 2.29) are supported unambiguously by two or more characters. Most characters show parallellisms or reversals. Not many groups are supported by more than one synapomorphy. Because of the amount of homoplasy, detailed discussion of every character and state is avoided. Characters that turned out to be uninformative for the analysis are not considered here [i.e., characters 9, 57, 63, 64, 83, 85, 87]. Also the characters pertaining to the stigma [76] and leaf base [81] were not used in the analysis, because I considered these too subjective. The characters concerning the shape of the bracts [37] and the shape of the calyx lobes [44] were left unused, because in former analyses they seemed to create more noise than pattern. The support of the larger groups of the tree is evaluated below. Branch numbers refer to the numbers (in bold type) in Figure 2.29.

Branch 63 supports the **clade of Butea and Meizotropis**. There are five synapomorphies supporting this branch, all other characters on this branch show parallellisms with one or more species in the other part of the tree. Synapomorphies are: the presence of indumentum on the upper surface of the leaves [19], on the standard [45], on the inside of the wing petal [53; reversal in *M. pellita*], and on the filament sheath [66; outside this clade only found in *S. acuminatus* as polymorphy], and the size of the anthers [67]. Since the flowers of *Butea*, and *Meizotropis* as well, are rather large, the last character is, considering the small size of the flowers of *Spatholobus*, not really comparable within the group under study.

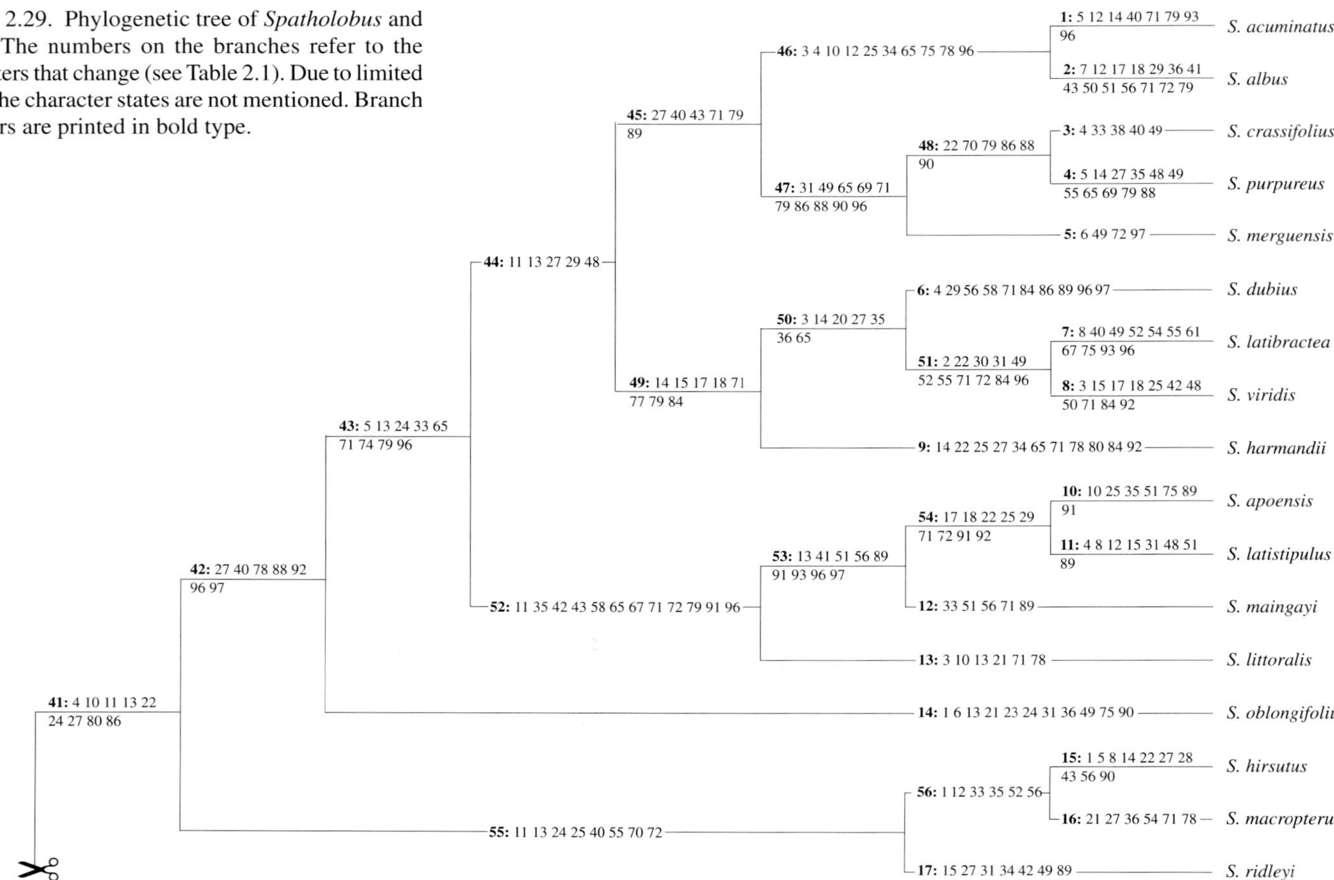

Figure 2.29. Phylogenetic tree of *Spatholobus* and allies. The numbers on the branches refer to the characters that change (see Table 2.1). Due to limited space the character states are not mentioned. Branch numbers are printed in bold type.

(Figure 2.29 continued)

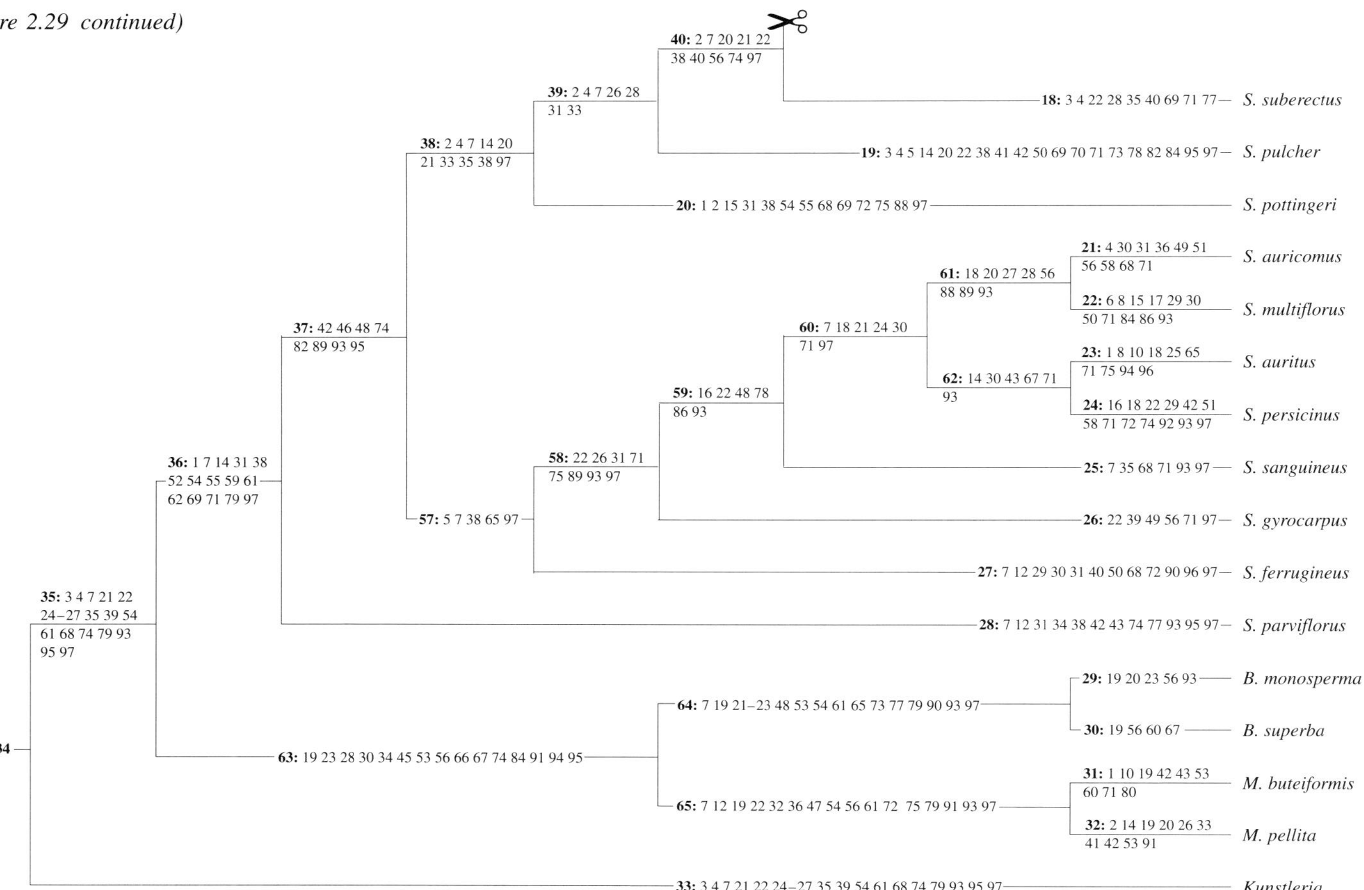

The other characters are the decurrent base of the leaflet [23; parallel in *S. oblongifolius*], the scalariform venation [28; parallel in *S. suberectus*], the secondary veins terminating in the margin [30; parallel in *S. ferrugineus*], the inflorescence with only first order side branches [34; parallel in the clade of *S. albus* and *S. acuminatus, S. harmandii, S. ridleyi* and polymorphic in *S. oblongifolius*], the absence of a dorsal auricle in the wing petal [56; parallel in several other species]; the medium position of the ovules in the ovary [74; parallel within the tree after node G, and the *S. ferrugineus*-clade], the curved anticlinal walls of the epidermis of the lower surface [84; parallel in *S. pulcher*, and polymorphic in *S. auricomus* and *S. ferrugineus*], the open ornamentation of the pollen grain [91; parallel with the *S. apoensis*-clade, and polymorphic in these species, and in *S. persicinus* and *S. latibractea*], the obtuse colpus ends [94; parallel with *S. pulcher*, and polymorphic in the *S. apoensis*-clade], the absence of mesocolpial pouches [95; parallel with *S. parviflorus* and *S. pulcher*].

The basal node [***branch 36***] of *Spatholobus* is supported by the following characters:
– the stem is not hollow [1], except for a reversal for the *S. macropterus*-clade;
– the wing petals are glabrous on the outside [52] as well as along the ventral margin [55], except for the *S. macropterus*-clade, *S. purpureus*, and *S. latibractea / S. viridis*;
– the outside and ventral margin of the keel petals are without indumentum [59, 62];
– the presence of usually 10 small nectary lobes [69]. *Spatholobus crassifolius, S. merguensis* and *S. pottingeri* show a reversal for the last character, and have no lobes, but only slightly thickened filament bases as in *Butea, Meizotropis* and *Kunstleria*.

Other characters are a synapomorphy for the genus *Spatholobus*, but change later in the rest of the tree (i.e. reversal or new apomorphies):
– a densely pubescent pulvinus [14], later becoming glabrous and reversing again to not densely pubescent;
– the number of pairs of nerves [31]; more than 8, many reversals in the rest of the tree to the plesiomorphic state of 5–8 pairs of nerves;
– the presence of two bracts at the base of the second order side branch of the inflorescence [38]; this is only the case in a few species (*S. crassifolius, S. parviflorus, S. pottingeri*, and *S. pulcher*), with reversals later to the plesiomorphic state of one bract at the base of the second order side branch;
– the indumentum on the ovary not very dense [71], later on larger parts of the tree reversal to plesiomorphic state of a densely hairy ovary.

Branch 37 supports the ***genus without S. parviflorus***. Four characters support this node with an apomorphic character state (i.e. a synapomorphy for *Spatholobus* from this point on, without *S. parviflorus*):
– the apex of the standard is emarginate [46];
– the base of the standard decurrent [48, with reversals later on];
– a bulbous septate hair base [82] is present;
– tannin is absent from the phloem in the main nerve, but present between phloem and xylem, or in the xylem itself [89].

A character state developed parallel with *Butea* is the footlayer/endexine ratio [93] being between 0.6 and 0.1; a few reversals to a thicker endexine occur.

A reversal to the presence of mesocolpial pouches [95] is also found at this point.

The **group of S. ferrugineus** on **branch 57** is supported by three apomorphic characters, all showing parallel developments in other groups: the bark of the branches is smooth [5; parallel with the top of the tree after *S. oblongifolius*, and with *S. hirsutus* and *S. pulcher*]; the filaments are connate only up to half of the total length of the filaments [65; parallel with the group after *S. oblongifolius* [branch 43]]. The character on the A/E-index [97] may have in this clade all three possible states.

The rest of the tree (from **branch 38** on) involves character states changing for, among others, several indumentum characters. With some exceptions, the species before branch 38 are more hairy than the species occurring in the top of the tree. The stem [2] is glabrous/glabrescent (with some exceptions, e.g., *S. pottingeri*), the pulvinus [14] is glabrous in most species (but reversals do occur), and the indumentum on the lower surface of the leaflets [20] is absent or nearly so (again reversals are found). The inflorescence [33], however, is more hairy (apomorphy) after branch 38 than before, and after *S. oblongifolius* the inflorescences are glabrous again (reversal). The presence of brachyblasts [35] is after branch 38 reversed to the plesiomorphic state (i.e. solitary flowers), but frequent changes to the apomorphic state (i.e. brachyblasts) occur. The shape of the leaflets [21] is in most species elliptic. From branch 38 on, no ovate leaflets are present, except for a few reversals. A few species, however, have obovate leaflets as well.

Branch 41 supports the second half of the tree. Reversals to the plesiomorphic states are the lateral leaflets [24] being symmetrical, except for *S. oblongifolius*, and the secondary venation [27] versus the venation as well being prominent at the lower surface of the leaflets; higher in the cladogram this character is changing to the apomorphic state of only a prominent midrib. Other changes are:

– the stipules with clearly visible nerves [10], a few reversals are present higher up;
– the stipules are glabrous [11], but only for the species up to the *S. maingayi*-clade; further on the other species are grouped by a reversal to an indument on the stipules;
– the colour of the pulvinus [13] shows the same pattern as the former character: only the species up to the *S. maingayi*-clade have a dark pulvinus;
– the apex of the leaflets [22] is now more strictly acute, although leaflets with acuminate apices still occur;
– the indumentum of the pod [80] is puberulous or nearly glabrous instead of densely pubescent; some reversals do occur;
– the presence of adaxial stomata [86] is a feature developed parallel with the upper part of the *S.ferrugineus*-clade.

The **group of S. macropterus** [branch 55] is supported unambiguously by the following character states:

- the midrib is flat or sunken on the upper surface of the leaflets [25]; this reversal is present in some other species as well;
- the bracteoles are placed immediately below the calyx cup [40], a plesiomorphic feature present in the largest part of the cladogram, except in the middle [branch 40–44];
- an indumentum on the ventral part of the wing petal [55; parallel in *S. purpureus, S. latibractea, S. viridis* and *S. ferrugineus*, in the last two only polymorphic]; this is a reversal also occurring in *Butea* and *Meizotropis*;
- an indument on the nectary [70]; this feature occurs in parallel in *S. crassifolius, S. pulcher* and *S. purpureus*;
- an indumentum halfway up the style [72]. It is an apomorphic state found in parallel in *S. ferrugineus, S. latibractea*, the *S. maingayi*-clade, *S. merguensis, S. pottingeri* and *S. viridis*. The majority of the species of *Spatholobus* have the indumentum up to the base of the style.

The ***S. maingayi-clade*** [branch 52] is supported by the following characters:
- brachyblasts present [35]; this is a reversal to an earlier apomorphic state in *Spatholobus*, brachyblasts are present in the majority of the species;
- a vexillary lobe that is not emarginate [42]; parallel with a few other species of *Spatholobus*;
- calyx lobes that are less than half the length of the calyx cup [43]; developed in parallel in *Spatholobus*, in *S. albus, S. hirsutus, S. auritus, S. persicinus*, although in these species they are perhaps not as extremely short as in the *S. maingayi*-clade;
- the absence of a lateral pocket on the keel [58]; parallel with *S. dubius, S. auritus, S. persicinus*; in all others the presence of pockets in both keel and wing petal is a plesiomorphic character;
- the anthers ranging in size from 0.5–1 mm [67]; parallel with *S. latibractea, S. auritus, S. persicinus*; the plesiomorphic state is less than 0.5 mm long;
- open ornamentation of the pollen wall [91]; although the clade is polymorphic for this character, it is the only clade with this feature within the genus, which also occurs as a parallelism in the *Butea*-clade.

The ***S. harmandii-group*** [branch 49] is supported unambiguously by five characters:
- a reversal to a pubescent pulvinus [14];
- the presence of more or less persistent stipellae [15]; parallel with *S. latistipulus, S. multiflorus, S. pottingeri, S. ridleyi*; although it is difficult, because in the outgroup *Kunstleria* stipellae are absent, this seems to be an apomorphy over the plesiomorphic early caducous stipellae.
- a glaucous upper surface of the leaves [17]; parallel with *S. apoensis, S. latistipulus, S. albus, S. multiflorus*, and polymorphic for *S. purpureus* and *S. merguensis*;
- and a brown lower surface of the leaves [18]; parallel with *S. apoensis, S. latistipulus, S. albus, S. auricomus, S. multiflorus*, and *S. persicinus*. These two states [17 and 18] are most probably related to each other;
- pod with a stipe [77]; parallelisms for *S. suberectus, S. parviflorus*, and *Butea*. Pods are not known from two species within the clade, *S. latibractea* and *S. viridis*;

– the wing of the pod is wider than the apex [79]; in this clade the apomorphic state of the whole genus is present versus the reversed state of the top of the tree after branch 43;
– the strongly undulated anticlinal cell walls [84]; this apomorphic state is present in a few other species, *S. ferrugineus*, *S. latistipulus* and *S. multiflorus*, which show this features occasionally.

Branch 45 connects the last two clades in the cladogram, those of *S. acuminatus* and *S. crassifolius*, and is supported by:

– a reversal to the state in which the bracteoles are placed immediately below the calyx cup [40]; *S. acuminatus* and *S. crassifolius*, however, have their bracteoles placed lower on the pedicel;
– the vexillary calyx lobe [43] is as long as the calyx cup; this is an apomorphic state shared with *S. parviflorus*;
– the indumentum of the ovary [71] reversed to the apomorphic state in the genus of a pubescent ovary; a reversal in some groups to a plesiomorphic densely hairy ovary took place earlier in the cladogram;
– the absence of the tannin (?) layer in the vascular bundle of the midrib [89]; this feature is shared with some other species: *S. apoensis*, *S. maingayi*, *S. ridleyi* and part of the *S. ferrugineus*-clade.

Summary of the character evaluation

In Table 2.3 an evaluation of all characters is presented. In order to avoid a long enumeration of characters, some characters are selected for a short discussion.

The species of *Spatholobus*, *Butea* and *Meizotropis* in the basal part of the cladogram (Fig. 2.29) are more hairy than *Kunstleria*. The species in the upper part of the cladogram are more glabrous than those of the lower part including *Kunstleria*.

In the second half of the cladogram: the shape of the leaflets [21] becomes elliptic, the apex of the leaflets [22] acute only, the lateral leaflets symmetric [24], with 5–8 pairs of nerves [31], brachyblasts are present [35], a well developed dorsal auricle on the keel petal is absent [56], the filaments are connate for less than half their length [65], the style remnant of the pod is bent downwards [or up; 78], adaxial stomata are present [86], transcurrent veins are absent [88], tannin is lacking in the vascular bundle [89], the infratectal layer is more granular [96], and the grains have a very small A/E-index [97]. Some of these character states are reversals, others synapomorphies.

There is a correlation between the size of the anthers [67] and pollen ornamentation [91]. Although the species with open ornamentation are polymorphic for this character, no other species in *Spatholobus* have this character state. In this analysis the species with the larger anthers are those with an ornamentation showing the sexine elements through the pores.

A very small A/E-index [97] is usually correlated with fused colpi, but this is not always the case, e.g., in *S. ferrugineus*, where the A/E-index is very small and no pollen grains with fused colpi were found.

Table 2.3. Character evaluation.

For each character the number of steps, the minimum/maximum number of possible steps, the retention index, the plesiomorphic state of the characters as optimized on the outgroup node, the apomorphic states, and remarks with possible transformation series of the states is given.

Abbreviations: (syn./aut.)apo. = (syn/aut)apomorphy, par. = parallelism, ples. = plesiomorphy, rev. = reversal, *B.* = *Butea*, *M.* = *Meizotropis*, *S.* = *Spatholobus*.

Character	Steps	Min/Max	RI	Plesiomorphic state	Apomorphic state	Remarks
1. Stem hollow	8	3–10	0.29	1 = yes	2 = no	synapo. *S.*
2. Indumentum	6	4–18	0.86	2 = pubescent / puberulous / sericeous	1 = sparse / not pubescent 3 = hirsute / strigose / pilose	2→3→1; glabrous from *S. pulcher* on.
3. Lenticels	7	2–9	0.29	2 = wart-like	1 = not conspicuous / small 3 = elongated	synapo. for *S., B., M.*; a few rev., a few par. to state 3.
4. Exudate	10	3–16	0.46	2 = absent	1 = present	synapo. for *S., B., M.*; later rev. in *S.*
5. Bark	8	3–15	0.58	2 = with ridges / wrinkles	1 = smooth	some par. within *S.*, after *S. oblongifolius* all state 1.
6. Ultrajugal part of the rachis	4	2–4	0.00	2 = ultrajugal part more than 1/10	1 = less than 1/10 of the rachis / no ultrajugal part	parallel in a few species.
7. Stipules	6	2–7	0.20	2 = not recurving	1 = recurving	synapo. *B.* and *S.*; early rev. in *S.* and also for *M.*
8. Stipules	5	1–5	0.00	1 = (early) caducous	2 = more or less persistent	a few parallellisms.
(9) Stipules symmetrical	1	1	0.00	1 = yes	2 = no (asymmetrical)	not informative.
10. Stipules nerves	6	1–12	0.55	2 = no visible nerves	1 = nerves visible	synapo. for *S.* after *S. suberectus*; *M. buteiformis* and *S. auritus* par.
11. Stipules indumentum	2	1–7	0.83	2 = hairy	1 = glabrous / glabrescent	synapo. after *S. macropterus*-clade; rev. after *S. maingayi*-clade.
12. Stipule length	7	2–9	0.29	2 = 2 or 3 times as long as wide	1 = as long as wide 3 = 4 or more times as long as wide	most species ples.; some either state 1 or state 3.
13. Pulvinus colur	4	2–8	0.67	2 = same colour as rachis	1 = black / darker than rachis	the same as character 11, some more par.
14. Pulvinus indumentum	10	2–20	0.56	2 = pubescent / hirsute	1 = glabrous / glabrescent / sparsely hairy 3 = densely pubescent / strigose	2→3→1; more glabrous after *S. pottingeri*; also rev.
15. Stipellae caducous	6	1–7	0.17	1 = (early) caducous	2 = more or less persistent	most ples., a few apo., including *S. harmandii*-clade.
16. Stipellae length	4	3–6	0.67	1 = up to the length of the petiolule	2 = conspicuously longer than the petiolule	a few apo., most within *S. ferrugineus*-clade.
17. Upper surface colour	7	3–9	0.33	1 = green / brownish (dried colour)	2 = greyish blue (glaucous)	most ples., a few apo., including *S. harmandii*-clade.
18. Lower surface colour	6	1–9	0.38	1 = green / brownish (dried colour)	2 = always brown	idem.
19. Upper surface indumentum	5	4–5	0.00	1 = glabrous to sparsely pubescent	2 = pubescent (dense) 3 = sericeous 4 = hirsute / strigose	synapo. for *B.* and *M.*
20. Lower surface indumentum	11	8–21	0.77	2 = pubescent (dense)	1 = glabrous to sparsely pubescent 3 = sericeous 4 = hirsute / strigose	2→4→1 or 2→3; synapo. after *S. pulcher* for state 1.

(*Table 2.3 continued*)

Character	Steps	Min/Max	RI	Plesiomorphic state	Apomorphic state	Remarks
21. Shape of top leaflet	10	4–15	0.45	1 = only elliptic	2 = ovate to elliptic 3 = obovate to elliptic	1→2→3→1; state 2 synapo. for *S.*, *B.* and *M.*, later to 3; rev. to 1 after *S. pulcher*.
22. Apex leaflet	17	8–24	0.44	2 = acuminate / cuspidate	1 = acute 3 = emarginate / rounded 4 = abruptly acuminate	many polymorphisms, state 1 after *S. suberectus*.
23. Base leaflet	6	5–6	0.00	1 = acute / obtuse / subcordate	2 = decurrent	only apo. for *B.*, *M.* and *S. oblongifolius*.
24. Lateral leaflet symmetrical	4	1–12	0.73	1 = more or less symmetrical	2 = strongly asymmetrical	synapo. for *S.*, *B.* and *M.*; rev. in *S.* after *S. suberectus* and in the *S. ferrugineus*-clade.
25. Main nerve	10	4–12	0.25	1 = sunken or flat	2 = raised	synapo. for *S.*, *B.* and *M.*; some rev. later.
26. Secondary nerves at upper surface	4	1–6	0.40	1 = flat	2 = raised	synapo. for *S.*, *B.* and *M.*; rev. from *S. pulcher* on and within the *S. ferrugineus*-clade.
27. Nerves at lower surface	9	3–19	0.63	2 = midrib and secondary nerves prominent	1 = only midrib prominent 3 = venation also prominent	2→3→2→1; synapo. for state 3 in *S.*, *B.* and *M.*, later in *S.* rev. to 2, and in top apo. state 1.
28. Venation	8	5–16	0.73	2 = reticulate-scalariform	3 = scalariform 1 = reticulate	2→3 and 2→1; state 3 synapo. for *S. suberectus*, *B.* and *M.*; state 1 synapo. after *S. pottingeri*, par. within the *S. ferrugineus*-clade.
29. Angle secondary nerves with midrib (average)	7	1–12	0.45	1 = up to 45°	2 = more than 45°	synapo. for top after *S. maingayi*-clade, a few par.
30. Secondary nerves terminate	5	2–10	0.63	1 = diffusely	2 = in the margin 3 = forming marginal arches	state 2 synapo. for *B.* and *M.*, par. with *S. ferrugineus*; state 3 a few species in *S.*
31. Pairs of secondary nerves (average)	9	1–15	0.43	1 = 5–8	2 = more than 8	apo. for *S.*, but rev. after *S. pottingeri*.
32. Place of inflorescence	1	1–2	1.00	1 = axillary / terminally	2 = terminally	synapo. for *M.*
33. Indumentum of inflorescence	9	5–10	0.20	1 = puberulous / pubescent	2 = hirsute / sericeous 3 = (nearly) glabrous	state 2 in part of *S.*, state 3 autapo. for *S. maingayi*.
34. Number of branches	7	4–11	0.57	2 = main axis with secondary and tertiary branches / fascicles	1 = main axis with (secondary) side branches / fascicles 3 = main axis with secondary, tertiary and quaternary branches / fascicles	state 2 for *B.* and *M.* and a few *S.*; state 3 for *S. parviflorus* and *S. ferrugineus* p. p.
35. Brachyblasts	13	5–16	0.27	2 = absent, flowers solitary	1 = present	synapo. for *S.*, *B.* and *M.*; later rev. after *S. pottingeri*.
36. Number of bracts of the first order (= secondary) side branch	6	1–9	0.38	2 = two bracts	1 = one bract	apo. for a few species in *S.*
(37) Shape of the bracts	11	4–11	0.00	2 = triangular	1 = linear 3 = broadly ovate 4 = elliptic, surrounding the bud	left out of analysis.

(Table 2.3 continued)

Character	Steps	Min/Max	RI	Plesiomorphic state	Apomorphic state	Remarks
38. Number of bracts of the tertiary branch / flower	4	1–4	0.00	1 = one bract	2 = two bracts	most ples., apo. only basal Asian species in *S.*
39. Length of the pedicel	3	2–3	0.00	1 = very short, up to 1 mm; flower nearly sessile	2 = pedicel > 1 mm; flower not nearly sessile	synapo. for *S.*, *B.* and *M.*; ples. *S. sanguineus* / *S. gyrocarpus*.
40. Place of the bracteoles	10	5–15	0.50	1 = immediately below the calyx	2 = on the upper half of the pedicel 3 = on the lower half of the pedicel	synapo. from *S. suberectus* on, state 3 only *S. acuminatus* and *S. latibractea*.
41. Indumentum of the calyx	4	2–6	0.50	2 = pubescent / sericeous	1 = sparsely pubescent / nearly glabrous 3 = hirsute	both apo. and par. in small groups.
42. Top of the vexillary lobe	13	7–16	0.33	equivocal	1 = emarginate 2 = not emarginate	only in *S.* state 1 ples.
43. Length of the vexillary lobe	9	4–16	0.58	2 = about half the length of the cup	1 = as long as the cup 3 = less than half the length of the cup	both apomorphic states present in but a few small groups.
(44) Shape of the other calyx lobes	12	3–13	0.10	3 = more or less rhomboid	1 = triangular 2 = truncate / somewhat rounded 4 = with a rounded apex	left out of analysis.
45. Indumentum on standard	1	1–4	1.00	2 = absent	1 = present	synapo. for *B.* and *M.*
46. Apex of the standard	1	1–6	1.00	2 = not emarginate	1 = emarginate	synapo. for *S.* after *S. parviflorus*.
47. Standard with auricles	3	3–4	1.00	2 = absent	1 = present	synapo. for *M.*
48. Base of the standard blade	11	5–18	0.54	2 = truncate	1 = decurrent into the claw	synapo. for *S.* after *S. parviflorus*; later rev. in tree and top *S. ferrugineus*-clade.
49. Dorsal auricle on wing	8	2–8	0.00	1 = present	2 = absent	par. in a few species of *S.*
50. Ventral auricle on wing	5	1–5	0.00	2 = absent	1 = present	idem.
51. Wing with lateral pocket	5	1–5	0.00	1 = present	2 = absent	idem.
52. Wing with indumentum outside	4	2–9	0.71	1 = present	2 = absent	synapo. for *S.*, with a few rev.
53. Wing with indumentum inside	2	1–3	0.50	2 = absent	1 = present	synapo. for *B.* and *M.*
54. Wing with indumentum on dorsal auricle/margin	5	1–6	0.20	1 = present	2 = absent	synpo. for all except *B.* and some species of *S.*
55. Wing with indumentum on ventral auricle/margin	7	3–13	0.60	1 = present	2 = absent	synpo. for *S.*, some rev. later.
56. Keel with a dorsal auricle	14	6–18	0.33	1 - present	2 = absent	par. in *B.*, *M.*, *S. pulcher* and top *S. ferrugineus*-clade.
(57) Keel with a ventral auricle				1 = present	2 = absent	left out of analysis.
58. Keel with a lateral pocket	4	1–7	0.50	1 = present	2 = absent	par. in a few species, including *S. maingayi*-clade.

(Table 2.3 continued)

Character	Steps	Min/Max	RI	Plesiomorphic state	Apomorphic state	Remarks
59. Keel with indumentum outside	2	2–6	1.00	1 = present	2 = absent	synpo. S.
60. Keel with indumentum inside	2	1–2	0.00	2 = absent	1 = present	only in B. superba and M. buteiformis.
61. Keel with indumentum on dorsal auricle/margin	3	1–4	0.33	1 = present	2 = absent	synapo. for all except B. and S. latibractea.
62. Keel with indumentum on ventral auricle/margin	1	1–5	1.00	1 = present	2 = absent	synpo. for S.
(64) Vexillary filament	1	1	0.00	2 = connate to the claw of the standard	1 = free	uninformative.
65. Filaments connate for	9	4–18	0.64	3 = more than half, but less than 3/4 of their length	1 = less than half of their length 2 = more than 3/4 of their length	par. after S. oblongifolius and the S. ferrugineus-clade.
66. Indumentum on filament sheath	2	2–5	1.00	1 = absent	2 = present	synapo. for B. and M.
67. Size of the anthers	5	3–11	0.75	1 = less than 0.5 mm	2 = more than 0.5 mm, but less than 1 mm 3 = more than 1 mm, but less than 2 mm 4 = more than 2 mm	par. in S. maingayi-clade and S. auritus, S. persicinus and S. latibractea; states 3 & 4 synapo. for B. and M.
68. Anthers	5	2–5	0.00	2 = fertile and alternately larger and smaller	1 = fertile and equal 3 = alternately fertile and sterile	state 1 synapo. for all; state 3 par. in a few basal species.
69. Nectary glands	6	3–11	0.63	4 = 10 thickened filament bases	1 = 10 2 = 5, all equal 3 = connate, no separate lobes	state 1 synapo. for S., state 2 par. in S. purpureus and S. pulcher, state 3 autapo. in S. suberectus.
70. Nectary with indum.	5	3–7	0.50	1 = absent	2 = present	par. in just a few species of S.
71. Indumentum of the ovary	15	3–18	0.20	2 = very densely pubescent / woolly	1 = pubescent / sericeous 3 = sparsely hairy	synapo. for S., with many rev., and some species with state 3.
72. Indumentum on the style	10	2–16	0.43	2 = at the base	1 = up to halfway 3 = up to at least 3/4	most ples., some state 1 or 3 par.
73. Number of ovules	2	2–3	1.00	1 = 2	2 = 2–4 3 = more than 4	state 2 autapo. for S. pulcher, state 3 synapo. for B.
74. Place of the ovules	6	2–8	0.33	1 = basal	2 = in the middle / apical	synapo. for B., M. and S.
75. Stipe of ovary present	8	1–13	0.42	2 = yes	1 = no	most ples., some par. state 1.
(76) Stigma	14	5–18	0.31	2 = capitate	1 = flat 3 = elongated	left out of analysis.
77. Stipe of pod exceeding calyx	6	3–8	0.40	2 = no	1 = yes	par. in B., S. parviflorus, S. suberectus and S. harmandii-clade.
78. Top of the pod	7	2–11	0.44	2 = with the style remnant pointing	3 = with the style remnant bent downwards 1 = with the style remnant bent upwards	2→3→1; synapo. from S. oblongifolius on, par. with top of S. ferrugineus-clade; par. in S. acuminatus-clade and S. macropterus state 1.

(Table 2.3 continued)

Character	Steps	Min/Max	RI	Plesiomorphic state	Apomorphic state	Remarks
79. Shape of the wing	5	1–8	0.43	1 = straight, as narrow as the seed-bearing part	2 = wider than the seed-bearing part	synapo. for all, but top tree and *M.* rev.
80. Indumentum of pod	3	1–12	0.82	2 = densely pubescent	1 = puberulous or nearly glabrous	synapo. for *S.* after *S. suberectus.*
(81) Hair base	15	14–19	0.31	1 = basal cell not in palisade layer	2 = basal cell in palisade layer	left out of analysis.
82. Bulbous septate hair base	3	2–6	0.75	1 = absent	2 = present	synapo. for *S.* / rev. for *S. pulcher.*
(83) Glandular hair	2	2	0.00	2 = stalk 2–5, head 4–8 cells	1 = stalk 1–2, head 2–6 cells; 3 = head cup-shaped	uninformative.
84. Epidermis lower surface	10	7–11	0.25	3 = undulate	1 = straight; 2 = curved cell walls; 4 = strongly undulate	state 2 par. in *B., M.* and *S. pulcher*; state 4 synapo. for *S. harmandii*-clade.
(85) Hypodermis	7	7	0.00	1 = absent; 2 = only near larger veins	3 = present throughout leaves	left out of analysis.
86. Stomata adaxial	9	5–17	0.67	1 = absent	2 = present	par. in *S.* after *S. suberectus* and in top *S. ferrugineus*-clade.
(87) Vascular bundle in midrib	5	5	0.00	1 = open	2 = closed / semi-closed	left out of analysis.
88. Veins transcurrent	12	9–19	0.70	2 = present	1 = absent (smaller veins palisade layer continuous)	synapo. for *S.* after *S. oblongifolius*, a few earlier par.
89. Tannin (?) containing cell layer between phloem and xylem	10	4–20	0.63	1 = present	2 = absent; 3 = in xylem itself	1→3→2; synapo. for *S.*, some rev; state 2 par. in top tree and *S. ferrugineus*-clade.
90. Tannin in bundle sheath	5	1–7	0.33	2 = absent	1 = present	par. in a few species of *S.*
91. Ornamentation pollen	13	12–13	0.00	1 = closed ornamentation (rugulate, psilate-perforate, fossulate)	2 = microreticulate / coarse perforate; 3 = verrucate	par. in *B., M., S. maingayi*-clade; *S. latibractea* and *S. persicinus* polymorphic.
92. Fused colpi	5	2–13	0.73	1 = absent	2 = yes, sometimes; 3 = often	synapo. after *S. oblongifolius*, some have state 3.
93. Foot layer / endexine ratio	8	2–10	0.25	1 = more than 0.6	2 = less than 0.6 and more than 0.1; 3 = less than 0.1	state 2 synapo. for *S.*, with some rev.; state 3 par. in *M.* and *S. parviflorus.*
94. Colpus ends	6	5–9	0.75	1 = acute	2 = obtuse	synapo. for *B., M.*, par. with *S. auritus.*
95. Mesocolpial pouches	3	1–6	0.60	1 = present	2 = absent	synapo. for *B., M., S. parviflorus*, par. with *S. auritus.*
96. Infratectal layer	32	30–34	0.50	2 = irregular (short) columellae	1 = well-spaced columellae; 3 = granules; 4 = indistinct / dense	many polymorphisms, after *S. oblongifolius* state 3 better represented.
97. A/E	11	2–17	0.40	3 = mean > 0.4	1 = mean < 0.28, max ≤ 0.31; 2 = 0.28 ≤ mean < 0.4	3→2→1; synapo. state 2 for all, with rev. to state 3 (among others *M.*); after *S. oblongifolius* state 1.

APPENDIX

For the study of the leaf anatomy, the following material was used
(from L, unless stated otherwise)

Butea monosperma: Backer 3424, 16773, 26695 (all Java), Chevalier 31338 (Vietnam, P), Dickason 6939 (Myanmar), Jacobs 4920 (Java), Kostermans 24393, 28597 (both Sri Lanka), Magnen, Gourgand, Châtillon s.n. (Cambodia), Teijsmann 1867 (Hort. Bog.), Thwaites 1465 (Sri Lanka, K).

Butea superba: Poilane 12288 (Vietnam), 14959 (Cambodia) (both P).

Meizotropis buteiformis: L herbarium sheet nr. 908.112-159, Griffith K.D. 1679 (East Bengal), Koelz 28394, 28408 (both India).

Kunstleria curtisii: KL 3428 (T & P 828) (Malay Peninsula).

Kunstleria forbesii: Elmer 13105 (Philippines, K), 21322 (Borneo), 21605 (Borneo), Forbes 3241 (Sumatra), Meijer 2256 (Borneo), SAN 37083, 39227, 89034, 89128 (all Borneo).

Kunstleria geesinkii: SAN 31574, 52944, 54866 (all Borneo, K).

Kunstleria kingii: FRI 13056 (Malay Peninsula, K), King's collector 3830 (Malay Peninsula, K).

Kunstleria philippinensis: BS 16144 (Philippines, US), Wenzel 818 (Philippines, US).

Kunstleria ridleyi: Elmer 21494 (Borneo), FRI 8360, 10518, 16919 (all Malay Peninsula), Hort. Bog. XVII.F.13, XVII.F.36, Kostermans 6569 (Borneo), Maingay 1168 (K.D.609) (Malay Peninsula), NBFD 4905 (Borneo), Niga Nangkat NN 193 (Borneo), S 15769 (Borneo).

Kunstleria sarawakensis: S 18036 (Borneo).

Spatholobus acuminatus: Lace 3033 (Myanmar, K), Langlassé 507 (Singapore, P), Maxwell 75-197 (Thailand), Wallich 5443 (India, K).

Spatholobus albus: S 23683, 32206, 32363, 32829 (all Borneo).

Spatholobus apoensis: Elmer 11795, 13300 (both Philippines).

Spatholobus auricomus: S 22446, 22517, 23063 (all Borneo).

Spatholobus crassifolius: Griffith 1826 (India, K).

Spatholobus dubius: Curtis s.n., April 1893 (Singapore, K).

Spatholobus ferrugineus: van Balgooy 3977 (Sulawesi), Blume s.n., 1284 (both Java), Boerlage s.n. (Java), Elsener H88 (Borneo), Hallier 1382 (Borneo), Maingay 530 (Malay Peninsula), Schiffner 2188 (Java), SF 40687 (Singapore), Teijsmann s.n. (Sumatra), de Vriese 66 (Borneo), Wiriadinata 3530, 3596 (both Borneo).

Spatholobus gyrocarpus: KL 2758, 3392 (both Malay Peninsula), Maxwell 86-530 (Thailand).

Spatholobus harmandii: BKF s.n. (Sangkhachand 238) (Thailand), F.C. How 72999 (Hainan, P), Lei 80 (Hainan), Maxwell 71-733 (Thailand), Pierre s.n., July 1866 (Vietnam), Poilane 11856 (Laos, P), 19757 (Vietnam, P), SF 23060 (Henderson) (Malay Peninsula, K), Tsang, Wai-Tak 288 (Hainan, P).

Spatholobus hirsutus: Endert 5162 (Borneo), Kato & Wiriadinata B.7001 (E Borneo), Kostermans 13652 (Borneo), Mogea 4085 (Borneo), Nieuwenhuis 948, 1079, 1440, 1444 (all Borneo), S 41420 (Borneo), SAN 76943 (Borneo).

Spatholobus latibractea: SAN 97466 (Borneo).

Spatholobus latistipulus: Elmer 21439 (Borneo), SAN 65877 (Borneo), Wiriadinata 3397 (Borneo).

Spatholobus littoralis: Backer 10099, 11016, 17061 (all Java), Blume s.n., 1723(?) (both Java), BS 28217, 30947 (both Philippines), Junghuhn 259 (Java), de Voogd 526 (Sumatra, BO), Winckel 325b, 488b (both Java).

Spatholobus macropterus: BRUN (Ashton) 3320 (Borneo), Elmer 21175 (Borneo), Forbes 3243 (Sumatra), FRI 10529 (Malay Peninsula), Kostermans 6872 (Borneo), Meijer 2508 (Borneo), PNH 23077 (Philippines), SAN 16421, 18841, 22567, 31293, 66164, 67684, 89836, 97576 (all Borneo).

Spatholobus maingayi: Elmer 17642 (Philippines), FRI 8373 (Malay Peninsula), PNH 2911 (Philippines), SAN 30594, 34322, 76900 (all Borneo), Santos 4064 (Philippines).

Spatholobus multiflorus: S 32370 (Borneo), SAN 91834, 96975, 110486 (all Borneo).

Spatholobus oblongifolius: Clemens 26354 (Borneo), Kostermans 13688 (Borneo), SAN 19832, 27454, 50481, 55708, 97669, 99420, 117192 (all Borneo).

Spatholobus parviflorus: Beusekom et al. 3607 (Thailand), BKF 37911 (Thailand), Hort. Bog. XVIII.D.23 (India), Kostermans 1171 (Thailand), Larsen et al. 31677 (Thailand), Maxwell 74-746 (Thailand), Murata T-17277 (Thailand), Pierre s.n. (Vietnam), Smitinand 4815 (Thailand).

Spatholobus persicinus: Endert 5410 (Borneo, BO), Kostermans 7108 (Borneo), Motley 734 (Borneo, K), Nooteboom 4374 (Borneo).

Spatholobus pottingeri: BKF 37401 (Thailand), Poilane 20366 (Laos, P), Thakur Rup Chand 6853a (India).

Spatholobus pulcher: Henry 12780B (China).

Spatholobus purpureus: Ridsdale 611 (India).

Spatholobus ridleyi: Forbes 3195 (Sumatra), Ridley 6401 (Singapore, K).

Spatholobus sanguineus: Elmer 17984, 18250 (both Philippines), Kostermans 21244 (Borneo), SAN 32504, 74457, 88852 (all Borneo), Wenzel 989 (Philippines).

Spatholobus suberectus: Poilane 15432 (Laos, P), 15870 (Laos, K), 25579 (Vietnam, P).

Spatholobus viridis: Kostermans 6586, 21654 (both Borneo), Meijer 2458 (Borneo, BO), SAN 97423, 110263 (both Borneo).

Chapter 3

OVERVIEW OF THE GEOLOGICAL HISTORY
OF SOUTHEAST ASIA

INTRODUCTION

In a historical biogeographical study the first question that comes to mind relates to the origin of the species. As De Jong (1987) stated: "What are the present-day ecological factors, and what are the historical ones? The range of a species can be changed actively by the organisms themselves or by external factors such as geographical or geological ones."

I think it is unsatisfactory to rely on the outcome of the analyses of area cladograms – based on the phylogeny and distribution of extant taxa – without some knowledge of the geology in the region. The area cladograms will probably indicate a common distribution pattern, but this does not explain the cause of that pattern. In historical biogeography the main assumption is that such a common pattern in distribution of different groups of organisms is related to geological or geographical (vicariant) events. These events may have caused isolation of species or an enlargement of the area. It is important, however, to specify the age and impact of the geological events before assigning them a role in the speciation of the groups studied. Whitmore (1981), e.g., states that, although the warm Tertiary climate influenced the flora and fauna in the Malay Archipelago, the most important climatic changes are those of the last two Ma. As an example of an ancient pattern, he cites the distribution of monsoon climate plant species, which have a very wide but disjunct distribution. This still reflects the more continuous range of distribution that was caused by a more seasonal climate in the now everwet parts in the past (for distribution patterns see also Van Steenis, 1961, 1979).

To get some idea about the geological history of Southeast Asia, it is necessary to study its fossil record, its palaeoclimatology, and its palaeogeography. A synthesis of the outcome of the cladistic analysis and the geological information may yield a reconstruction of the historical development of the distribution of the present day taxa.

This overview is based on literature, which is cited in the text. The part on the geology is a summary (and synthesis) of several articles and books of Hamilton (1973, 1979, 1988), Hutchison (1989a, 1989b, 1992, 1996), and Metcalfe (1988, 1990, 1991, 1994a, 1994b, 1996). These are mostly not cited individually to avoid a text unreadable by the amount of citations. Additional literature is cited separately. The overview presented here is restricted to the region of occurrence of the genera *Spatholobus*, *Butea*, *Meizotropis* and *Kunstleria*: Southeast Asia (from the Himalayas and Yunnan to southern India and Indochina), and the western Malesian Archipelago (including Sulawesi and the Philippines) (Fig. 3.1). In Table 3.1 some of the characteristics of each period and some of the tectonic events are listed.

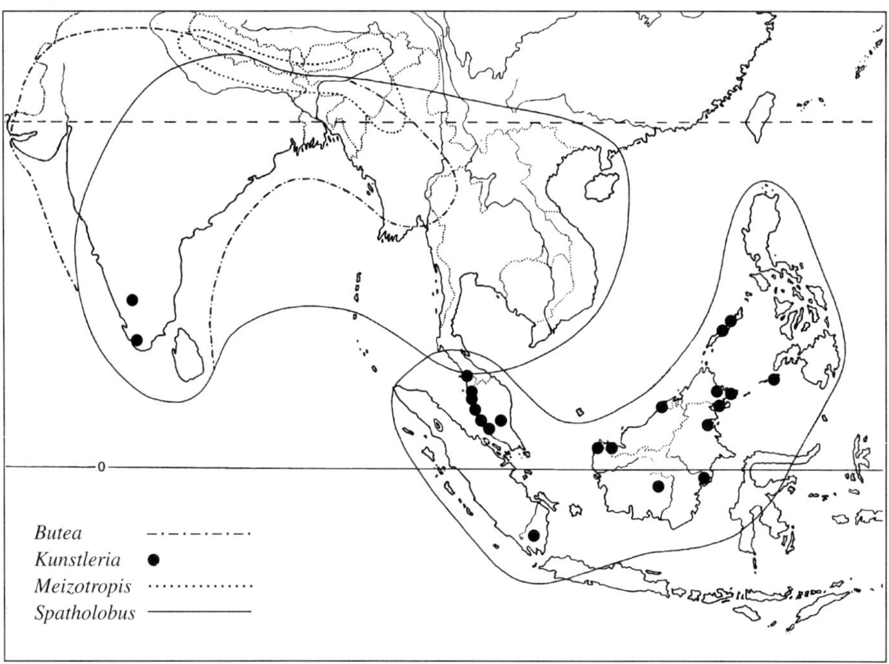

Figure 3.1. Map of Southeast Asia and the West Malesian Archipelago showing the distribution of the genera *Butea, Kunstleria, Meizotropis,* and *Spatholobus.*

FOSSIL EVIDENCE OF LEGUMINOSAE IN SOUTHEAST ASIA

The first angiosperms probably appeared in the Southeast Asian region in the Early Cretaceous (Wing & Sues, 1992). They already had undergone significant diversification in the lower latitudes, but were generally still rare or absent in mid- and higher latitudes. During the Late Cretaceous, the number of angiosperms increased considerably, and by the end of that epoch 50% to 80% of the vascular plant species were flowering plants. In the opinion of Wing and Sues the early angiosperms probably were weedy plants or small trees that flourished in disturbed settings or undergrowth of forests dominated by large coniferous trees. During the Early Eocene angiosperm-dominated forest was at the height of its distribution, subtropical forest may have reached up to 60° N and tropical rain forest occurred as far north as 30° N.

On account of the presumed unlikelihood of long dispersal occurring in many disjunct genera, Walker (1982) states that angiosperms must already have been in existence since the Late Jurassic (135 Ma). However, there is no fossil evidence for this. He suggests that *Kingiodendron* (Leguminosae), a genus with a very disjunct distribution from India over the Philippines into Fiji, might be an example of this.

Raven and Polhill (1981) mention the Maastrichtian (Upper Cretaceous) as being where the first records of the Leguminosae (both pollen and macrofossils) were found. The Upper Cretaceous records of fossil wood and pollen are regarded as unreliable by

Table 3.1. Schematic overview of geological events in relation to Southeast Asia.

Time	Events	Sea level / Climate	Fossils / Flora
Palaeozoic (600–225 Ma).	SE Asia part of Gondwanaland (northern Australia).		
Silurian /Ordovician or Carboniferous (Upper lower Palaeozoic, 500–400 Ma or up to 280 Ma).	Blocks (Yangzi, Phu Hoat, North China and South China (Metcalfe), Indochina/ Eastern Malaya, West Borneo Basement) rifting away from Gondwana in sets. *Burrett: this event in Cambrian (600–500 Ma).*		
Early Carboniferous (340 Ma).	South China welded with Indochina/ East Malaya to form Cathaysialand. Blocks are in equatorial part. Still close to Gondwana are Lhasa, Himalaya, Shillong, Indian platform, Burma Plate, Sinoburmalaya (Sibumasu), Shan-Tai.		Cathaysian flora (not glacial) for the equatorial blocks, the other blocks remained under influence of Gondwanaland in Carbo-Permian times, including glaciation.
Early Permian (270 Ma).	Rifting of Cimmerian continent off Gondwana (among others Sibumasu).		
Late Permian/Early Triassic (240–220 Ma).	Collision of Sinoburmalaya (Sibumasu) with Asia (and Indosinia). North China welded with Cathaysia.	Up to now Eurasian Plate marine.	
Triassic (225–195 Ma).	Burman Plate present as island arc in front of future Southeast Asia.		
Early Jurassic (195 Ma).	*Acc. to Audley-Charles, rifting of Tibet, Burma, Thailand, Malaya, Sumatra.* Opening of Proto-South China Sea and rifting of West Borneo from Indosinia. Separation of India from Gondwanaland.	Eurasian Plate above sea level.	
Mid-Jurassic.	Western Burman Plate collided with Asian Plate. Closure of Tethys Ocean. East margin of China volcanic arc up to Indosinia.		

(continued on next pages)

(Table 3.1 continued)

Time	Events	Sea level / Climate	Fossils / Flora
Late Jurassic.	*Acc. to Audley-Charles collision of blocks mentioned above with Asia.*		
Cretaceous (135–65 Ma).	Burman Plate was island arc on the Asian margin. *Burrett 1990: South Sumatra and South Kalimantan still close to Australia.*		
Late Cretaceous.	Collision of Burman Plate with Asia (Shan states). Arrival of Woyla terranes Sumatra.	Sea level up to +350 m.	First records of Leguminosae (Raven & Polhill, 1981) unreliable records for India and China.
Early Tertiary (65 Ma onwards).	Oceanic crust Philippine Sea Basin began forming at the southern boundary of the Pacific Plate.		
Palaeocene (65–54 Ma).	Opening of the Makassar Basin; West Sulawesi rifting off SE Borneo. Deltas (Baram, Tarakan) develop. Formation of sediments of Northwest Borneo Geosyncline (Rajang group). Borneo south and southeast of Rajang group form landmass together with West Sulawesi and Java Sea.	Tropical monsoon. Episodes of lower sea levels.	Fossils indicating presence of Caesalpinioideae and Mimosoideae (Herendeen et al. 1992).
Early Eocene (53 Ma).	Indian collision and subduction below Burma (uplift of Indo-Burman Ranges). Spreading after separation of Australia from Antarctica. Start of subduction of Australian Plate in Sunda Trench.		
Middle Eocene.	*Burrett, 1990: collision of S Sumatra and S Kalimantan with SE Asia.*	Tropical monsoon episodes of lower sea level.	All three subfamilies are represented in fossil record.
Late Eocene (45 Ma).	Final collision Indian Plate with Eurasia (Himalaya). Philippine Sea Basin rotate clockwise and drift northward.		

(*Table 3.1 continued*)

Time	Events	Sea level / Climate	Fossils / Flora
Oligocene (38–26 Ma).	Opening of the South China Sea (32–17 Ma); rifting of blocks towards Borneo from Chinese margin. South China Sea. Start of rift of fragments of Chinese continental margin.	Wet tropical monsoon or tropical rain forest. Sea level fall of –250 m (Late Oligocene).	Legume species known from Yunnan and India.
Miocene (26–7 Ma).	Java becomes part of the active plate margin. Spreading in Philippine Sea Basin ceased 26 Ma.	Tropical monsoon.	Fossils identified up to generic level in China and India. Migration from Legumes into India from Africa and Southeast Asia well documented (Awashti, 1992). Legumes dominant in India.
Middle Miocene.	Opening of Andaman Sea (13 Ma) as result of the rotation of the Malay Peninsula. End of spreading activity in South China Sea by collision of Palawan with West Philippines block. Volcanic arc in N Sulawesi active (east facing before rotating later). Collision of Banggai-Sula with Sulawesi Trench.	Mid-Miocene sea level changes in three stages to +220 m. Tropical monsoon.	
Late Miocene-Pliocene.	Volcanic arc active in Philippines. Uplift of Meratus Mts and Semitau Range (other mountains Cretaceous). Collision of Australian continental shelf with Timor.		
Pliocene (7–2 Ma).		Sea level rise to +140 m.	Fossil record more scanty (drier period/lack of studies).
Pleistocene.		Many glacial periods.	

Herendeen et al. (1992), because they cannot unequivocally be assigned to the Leguminosae (see also Wheeler & Baas, 1992). The Palaeocene fossils (pollen as well as fruit and wood) are more informative, however, and suggest the presence of Caesalpinioideae and Mimosoideae. In the Eocene both subfamilies were abundant. The origin of the Papilionoideae is unknown, but by the Middle Eocene they were diverse. From the Middle Eocene onwards all three subfamilies are represented.

Modern legume groups originated by the end of the Cretaceous, when Eurasia was still in contact with Africa, and there were still overland connections between North America and Europe (Raven & Polhill, 1981). These authors suggest that Africa has been a prime area for the early radiation and evolution of Leguminosae, based on both the geographical relationships and the common occurrence of many legumes in moist and dry forests in Africa. They explain the relatively poor fossil record of Eurasia and the less abundant current occurrence in tropical Asia (as opposed to that of Africa, Madagascar and Latin America) by the existence of a post-Palaeocene water barrier between Africa and Eurasia. Later, when exchange was possible, difficulties in penetrating forests containing competitive groups may have prohibited further migration.

Probably due to climatic factors several taxa were more widespread than they are today. The limited fossil record in New Zealand, Australia, China and India, however, may indicate that these areas were reached relatively late (Herendeen et al., 1992; Polhill & Raven, 1981). Fossil data as well as data from extant taxa indicate that the family has had a long evolutionary history in tropical America and Africa/Madagascar. Despite water barriers, there has been a floristic exchange between Africa/Madagascar and the Americas, between Africa and Europe, and between South and North America.

A tropical North American origin has been assigned to the Papilionoideae by Lavin (1995). During the Early Tertiary they could have spread out over North America and Eurasia before both continents were separated. The latter suggestion, also known as the Boreotropics hypothesis (Lavin & Luckow, 1993), more readily explains the occurrence of relatives in Southeast Asia, Africa and South America than by direct exchanges (e.g., between South America and Africa) (Fig. 3.2). In addition, the hypothesis satisfactorily explains the occurrence of fossils in southeastern North America, whose closest relatives are living in tropical South America (Herendeen et al., 1992). The disjunct occurrence of the closely allied Southeast Asian genera *Derris* and "American *Derris*" (*lonchocarpus*) (Geesink, 1984) may be explained by this theory as well.

In China and India reliable fossil records of the Leguminosae extend as far back as the Eocene (Guo & Zhou, 1992; Awasthi, 1992); from the Late Cretaceous there are a few unreliable records only from both China and India. There are two legume species known from the Oligocene of Yunnan. Leguminosae are well represented, however, in the Miocene fossil record throughout China, and have often been identified up to generic level (e.g., *Dalbergia, Desmodium, Gleditsia, Pueraria, Sophora*). Probably due to a lack of comprehensive studies little is known from the Pliocene and Quaternary of China.

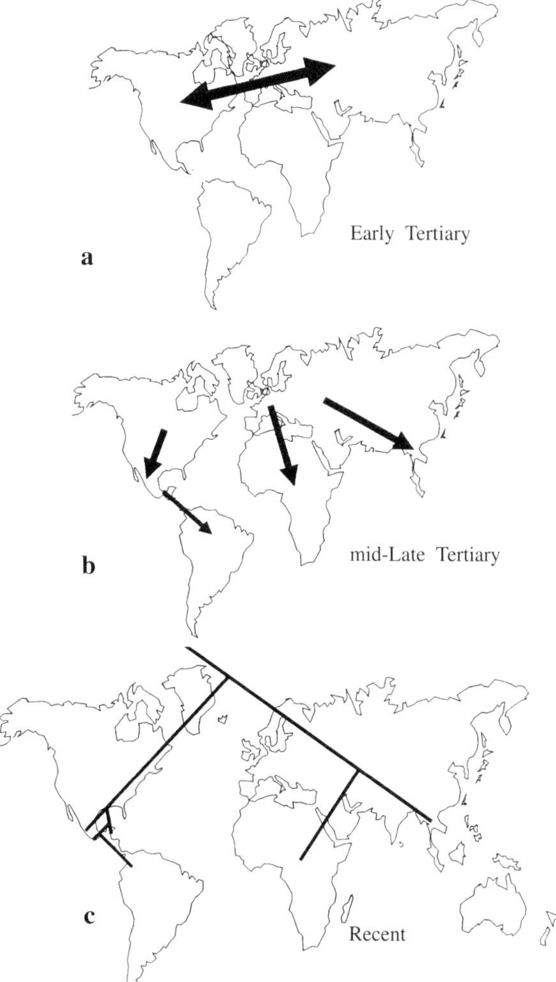

Figure 3.2. Maps illustrating the Boreotropics hypothesis (after Lavin, 1995).

a: Arrow suggesting direct interchange of tropical biotas during the Early Tertiary.

b: Arrows suggesting movement of Early Tertiary northern tropical elements with the onset of mid-Tertiary climatic cooling and separation of the North Atlantic.

c: 'Area cladogram' of present-day remnants of Early Tertiary north tropical groups predicted by the Boreotropics hypothesis.

In northeastern India legumes are present in the fossil flora since the Late Palaeocene. During the Oligocene and Early Miocene they are also found throughout peninsular India. Migration of legumes into India from Africa (indirect) and Southeast Asia, and *vice versa*, is well documented since the Miocene. By the Middle Miocene legumes were dominant in India. They reached a maximum diversity and were one of the major components of the tropical evergreen forest. By the end of the Pliocene, as the climate became drier, the tropical evergreen forest diminished and many taxa became extinct. Some taxa survived, however, in the evergreen forests that remained in the Western Ghats and northeastern India. The deciduous legumes living today in India may be regarded as the result of adaptations to a drier climate.

CLIMATE

During the Mesozoic-Early Cenozoic, annual rainfall was relatively high at the mid-latitudes. The equatorial region, however, was drier, and characterised by higher temperatures and light levels (Wing & Sues, 1992). The temperature reached peak values during the Early Eocene. A cooling and drying trend followed during the mid-Oligocene transgression. From the Early Miocene on the climate was warmer again, but the overall trend during the Late Cenozoic is a cooler and drier climate, although this has alternated with warmer temperatures. A decline in temperature after the mid-Miocene (13 Ma) is associated with a retreat of the palaeotropic flora and expansion of deciduous forests and grasslands (Potts & Behrensmeyer, 1992).

During the Tertiary and Quaternary, climatic pulses of temperature resulted in an expansion of new unexplored environments. This led to an increase in the diversity of Leguminosae from drier environments (with many herbaceous Papilionoideae) to tropical regions (rich in Caesalpinioideae, Mimosoideae and woody Papilionoideae). Processes that indirectly affected the climate are plate tectonics, mountain building, glaciation and sea level changes, and volcanism. The tropical region became progressively warmer and drier during the Miocene (Axelrod, 1992), resulting in the growth of savannahs and grassland in which tropical legume species could adapt to drier conditions.

During the periods with a drier climate, the evergreen forests were restricted to only certain parts of India. The species were forced to adapt to drier conditions or disappear (Awasthi, 1992). The South China Sea region, Borneo, and Brunei became more seasonal (Morley & Flenley, 1987; Van Steenis, 1961). During the Pleistocene these authors suggest a 'savannah corridor' connected the Malay Peninsula and Sumatra with South Borneo (Fig. 3.3). In this reconstruction the northern part of Borneo remains everwet. The temperature was lower and resulted in a forest limit that was about 1700 m below the present one. The Pleistocene glaciations resulted in eustatic low sea levels that profoundly changed the shape of the Malay Archipelago. The existence of a 'savannah corridor' was mentioned by Van Steenis (1961) in connection with the now disjunct areas of several drought plants. Verstappen (1980) adds that drier conditions were also inferred from fossils of grazing animals found on Java and other islands. The savannah corridor was bordered by monsoon tropical forest during periods of low sea level (Adams, 1995). Van der Kaars and Dam (1995) provide data from the Bandung area covering the last 135,000 years. They found dry conditions around 135,000 years B.P., warm and humid ones during the interglacial between 126,000 and 81,000 years B.P.; the climate became dry again between 81,000 and 74,000 yr B.P., but later it became more wet. The last drier and cooler period was between 47,000 and 20,000 yr B.P, during the Last Glacial Maximum.

Southwest China had a tropical monsoon climate during the Palaeocene, as indicated by the presence of megafossils (Guo, 1993). The climate was a little drier during the Eocene, but tropical monsoon forests were still present. During the Oligocene the climate was warmer and more humid, characterised by a wet tropical monsoon or tropical rain climate. Towards the end of the Oligocene climatic conditions were drier, and at the beginning of the Miocene also cooler. This change may be the result of uplifted

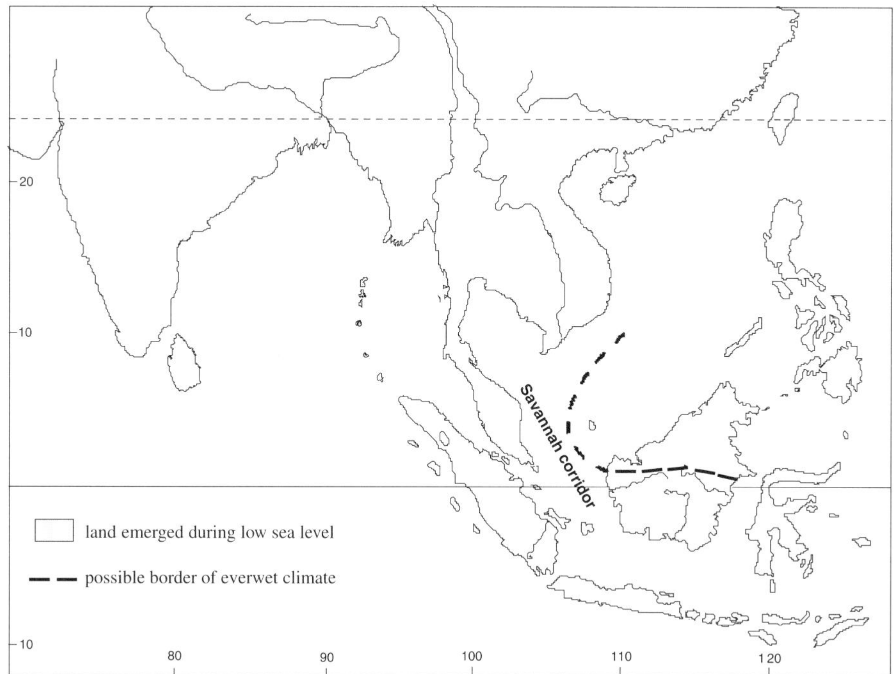

Figure 3.3. Map illustrating the Malesian Archipelago during low sea level. Modified after Morley & Flenley (1987) and Verstappen (1975).

plateaus (Himalayas and others). On plains and hills a tropical monsoon climate prevailed, with the Leguminosae as one of the dominant families. In the Pliocene a montane tropical monsoon climate was present in NW Yunnan and to the north in S Sichuan.

According to Ferguson (1993) the climate in northern China was modified by the rapid uplift since the Pleistocene of the mountain plateaus that threw a rainshadow over the area. It became more extreme: colder and drier during the glacials, warmer and moister during the interglacials. Even in the south and southwest of China this was felt, but it was less pronounced. Yunnan had a stable flora and fauna up to the Pleistocene, and was later affected by the influence of a drier climate.

EUSTATIC SEA LEVEL CHANGES

Eustatic sea level variations have a profound impact on the environment and are particularly important for the West Malesian area. A lowering of the sea level (Table 3.1) enabled a periodic exchange between regions that were otherwise isolated.

From the Jurassic to the Cretaceous large parts of South Tibet, Indochina, and probably Sumatra were exposed (Audley-Charles, 1987). During the late Cretaceous and Palaeocene (85–55 Ma), sea levels reached an all time high of 350 m above the present level (Hutchison, 1989a; Dercourt et al., 1993). It stayed high until the Late Oligocene except for some episodes of lower sea levels during the Late Palaeocene, and the Lower and Upper Eocene. During the Late Oligocene a spectacular fall in sea level to 250 m below the present level exposed the entire landmass of Sundaland. From then on the sea level rose again up to 220 m above the present level in mid-Miocene times, and the Indian continent and the Himalayas (before the real uplift) were only joined by an isthmus (Cavelier et al., 1993). This was followed by a drop in sea level that occurred in three phases to 220 m below its present level from the Middle to the Late Miocene. Again the Sundaland Plateau was exposed. At the beginning of the Pliocene a rise to 140 m above present-day sea level occurred. During the Pleistocene, changes in sea level were more frequent, but less spectacular.

During the Pleistocene, glacial periods occurred for longer periods than the interglacials (Morley & Flenley, 1987). There is a coincidence between climatic changes and the rise and fall of the sea level. On the continent an increase in drier conditions occurs when the sea level is low, and presently inundated parts of Sundaland become inhabited by landplants from the now drier continent. The newly formed landmass created a corridor with savannah vegetation in its central part, and monsoon forest along the edges. Land bridges allowed plants and animals to migrate from Thailand to Sulawesi. The existence of a savannah corridor (Fig. 3.3) was already mentioned by Van Steenis (1961, 1979) in his articles on vegetation types in the Malesian Archipelago. He suggested that during the Pleistocene Ice Ages the seasonal monsoon climate zones were extended, providing a pathway from the Southeast Asian mainland into the Malesian Archipelago (and even into Australia).

Plate tectonic reconstructions often do not adequately demonstrate the land–sea distribution. Maps that show clear coastal outlines are important for biogeographical reasons, however. Colour maps with indications which areas have been above sea level in certain periods have been published by, e.g., Rangin et al. (1990b) and Dercourt

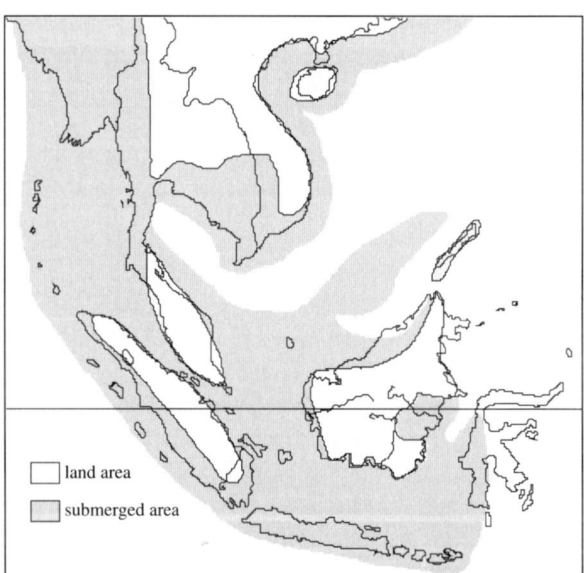

Figure 3.4. Map of the emerged parts of Southeast Asia during high sea levels in Middle Eocene to Late Miocene times. Modified and simplified after Rangin et al. (1990b).

et al. (1993). The reconstructions of Rangin (Eocene to Present) show the Malesian area during high sea level, and it is remarkable to see that the area is divided into a few larger emergent landmasses: Borneo, the Malay Peninsula, Sumatra (in part) and West Java. Figure 3.4 shows the emergent landmasses during high sea levels in Middle Eocene to Late Miocene times. In the Pliocene the emergent parts are similar to the mountain ranges presently occurring, including the mountains west of Phnom Penh.

GEOLOGY

Southeast Asia represents a complex region that is influenced by the convergence of four major lithospheric plates: The Eurasian Plate, the Indo-Australian Plate, the Pacific Plate and the smaller Philippine Sea Plate. The region is further complicated by the presence of numerous smaller continental fragments and island arcs that are caught between these plates (Fig. 3.5).

The geology of the region has been summarised in comprehensive studies by Hamilton (1973, 1979, 1988), Hutchison (1989a, 1989b, 1992, 1996), and Metcalfe (1988, 1990, 1991, 1994a, 1994b, 1996). Geological evidence on which this overview is based can be found in these references. Additional references are mentioned where necessary in the text. Biogeographical studies of the East Malesian Archipelago, including an overview of the geological history with emphasis on New Guinea, can be found in De Boer (1995), Turner (1995), and Vermeulen (1993). De Jong and Treadaway (1994) discussed the biogeography and geology of the Philippines.

Tectonic mechanisms

Although there were scientists (Lam, 1930; and references cited by him) who adopted the theories of continental drift at an early stage, it was not until the sixties that the theories of Alfred Wegener (1915) became widely accepted. Since then a wealth of geological information has been acquired using highly specialised disciplines such as palaeomagnetics, seismology, isotope stratigraphy and palaeontology. This has contributed to the elaboration of the plate tectonic concept.

The surface of the earth is covered by lithospheric plates that may be continental or oceanic (or both) in origin. Plates move relative to one another along plate boundaries (Fig. 3.5). Plate boundaries are convergent (active) when one plate is moving below the other (subduction). Divergent plate boundaries (the mid-oceanic spreading ridges) are located at the position where two plates move apart.

The subduction of one plate below another results in characteristic geological features more extensively explained in Hamilton (1988) and Daly et al. (1991); some are shown in Figure 3.6. At convergent margins, sediments may be scraped off the subducting plate and welded against the overlaying margin where they form an accretionary wedge. At the zone where the plate is subducted to depths greater than 600 km, the high temperature causes the sinking slab to melt. Volcanic arcs develop on top of the overriding plate. Between the accretionary wedge and the volcanic arc the fore-arc basin is situated, usually a sub-marine basin that originated due to extension of the overriding plate. At the other side of the volcanic arc a back-arc basin may be present.

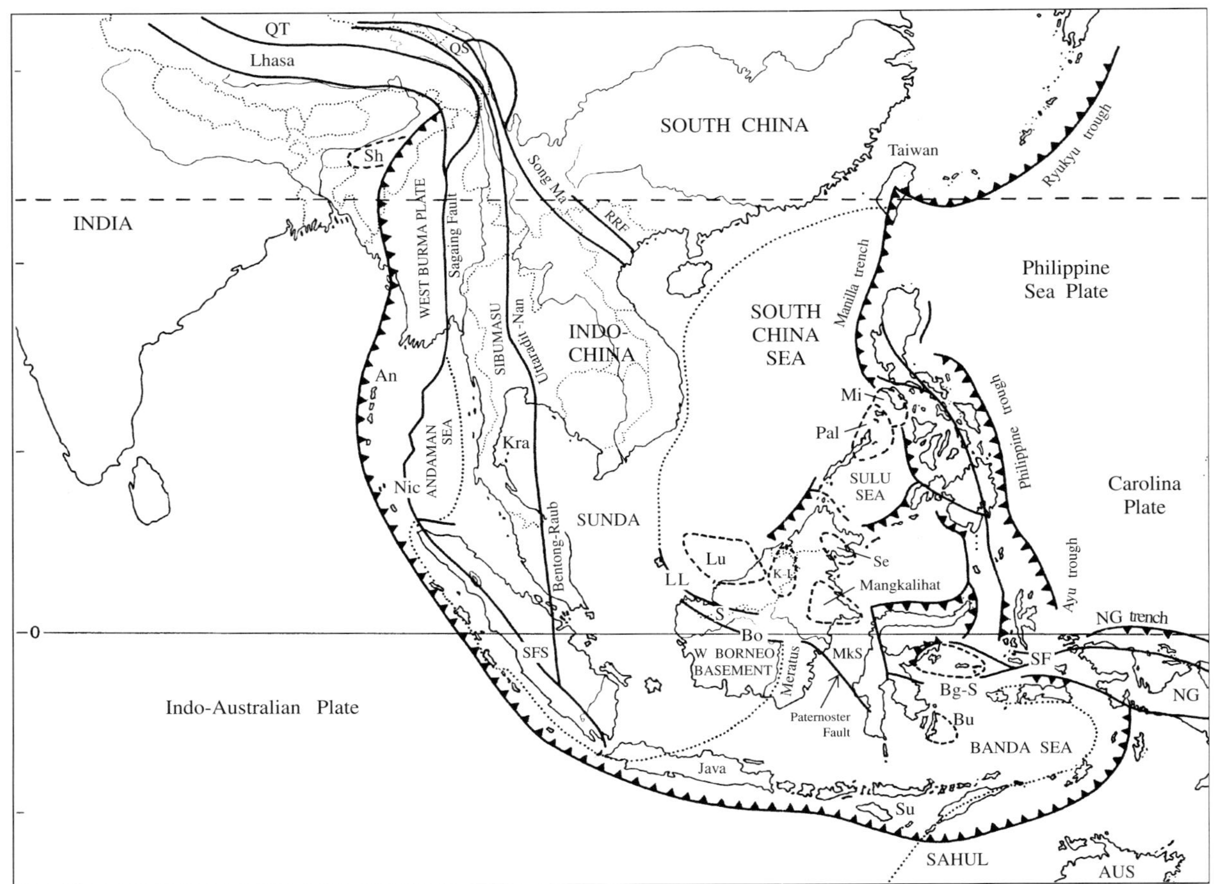

Figure 3.5. Map of SE Asia showing the main geological features. Heavy lines represent fault systems, indented lines show plate margins with the indentation on the overriding plate, and dotted (bold) lines represent the border of Sahul and Sunda platform.

An — Andaman Islands
AUS — Australia
Bg-S — Banggai-Sula
K-L — Kelabit-Longbowan
Kra — Isthmus of Kra
LL — Lupar Line
Lu — Luconia
Mg — Mangkalihat
Mi — Mindoro
MkS — Makassar Strait
NG — New Guinea / Irian Jaya
Nic — Nicobar Islands
Pal — Palawan
QS — Qamdo Simao terrane
QT — Qiangtang terrane
RRF — Red River Fault
S — Semitau
Se — Segama
SF — Sorong Fault
SFS — Sumatra Fault System
Sh — Shillong Plateau
Su — Sumba

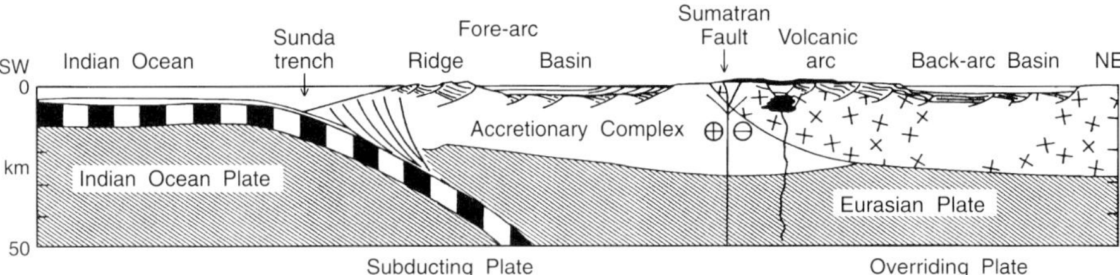

Figure 3.6. Schematic cross section of a subduction zone (after Simandjuntak & Barber, 1996)..

Plates do not always approach each other in a perpendicular fashion. Oblique subduction or strike-slip movement between plates are more common, and may result in large faults (e.g., the Sumatran Fault). The rigid subducted slab is reflected by a high density of earthquakes known as the Benioff Zone.

During their origin, rocks may become magnetised by the earth's magnetic field and may then receive a polarisation corresponding to the direction of the earth magnetic field at that time. The earths magnetic field is considered to be a dipole field, the axis of which has an average position parallel to the rotation axis of the earth. Because continents may have moved since, palaeomagnetism can give valuable information on their original latitudinal position.

Present situation

The Southeast Asian part of the Eurasian Plate represents an assemblage of larger and smaller continental fragments that probably originated from Gondwanaland. The region can be divided into several structural units: India and the Himalayas, West Burma, Sinoburmalaya (Hutchison) or Sibumasu (Metcalfe), South China, Indochina and East Malaya, and several smaller fragments (Fig. 3.5).

India is part of the Indo-Australian Plate and collided with the Eurasian Plate during the Late Eocene resulting in the formation of the Himalaya mountains. Early collision events during the Early Eocene between India and Burma caused the uplift of the Indo-Burman Ranges. The West Burman Plate represents a narrow plate extending from the Indo-Burman Ranges to the Shan and Sagaing Faults (Fig. 3.5, 3.13).

Sundaland itself is composed of a complex of cratonic terranes (= structural units of the earth's crust) and younger orogenic belts that developed since the Late Triassic. There is no seismic activity except along some Cenozoic faults (e.g. the Red River Fault). Sundaland is bordered by the Shan Boundary and Sagaing Fault in Burma, and its outlines continue southwards along the Andaman–Nicobar Ridge and offshore to the west of Sumatra. On West Java it extends eastwards and continues (in a fragmented way) to the Meratus Mountains in SE Borneo. West Sulawesi represents a piece of Sundaland that drifted away from the southeastern margin of Borneo in the Early Tertiary opening up the Makassar Strait. On Borneo, the margin of Sundaland can be traced along the Lupar Line, east of Kuching, and extends into the South China Sea along the Natuna Ridge, and off the coast of Vietnam towards the Song Ma Suture. The Song Ma Suture represents the area where the blocks of South China and Indochina are welded. The Sunda shelf is the extension below sea level of the continent, connecting Borneo, Java, and Sumatra. The maximum depth of the continental shelf is 200 m.

The greater part of the Philippines is composed of a separate block which is enclosed between the Pacific and Eurasian Plates. In the west, the South China Sea Basin (Eurasian Plate) is subducting below the Philippines along the Manila Trench. In the east, the Pacific Plate is subducted along the Philippine Trench. The Philippines consist partly of continental fragments that were separated from China after opening of the South China Sea Basin.

Australia, including the Sahul shelf and New Guinea, is presently colliding with Indonesia. Fragments of this shelf are found in several parts of the East Malesian Archipelago.

The major rifting events

There is much evidence that most Southeast Asian terranes formed an integral part of Gondwanaland in the Palaeozoic. I follow Metcalfe, who summarised the dispersal of the fragments as three separate rifting events:

- (Early) Palaeozoic rifting of, among others, N China, S China, Indochina and E Malaya, and the Tarim terrane (Fig. 3.7).
- Early to Middle Permian rifting of Sibumasu, Qiantang, Iran and Turkey (together forming the Cimmerian continent). The continental sliver disintegrated, after which Sibumasu, Qiantang, Iran and Turkey travelled rapidly to the north during the Permo-Triassic (Fig. 3.8).
- Late Triassic rifting of Lhasa to Late Jurassic rifting of W Burma, Woyla (Sikuleh and Natal in Sumatra), and Mangkalihat (part of E Borneo), W Sulawesi and Banda allochton (Fig. 3.9).

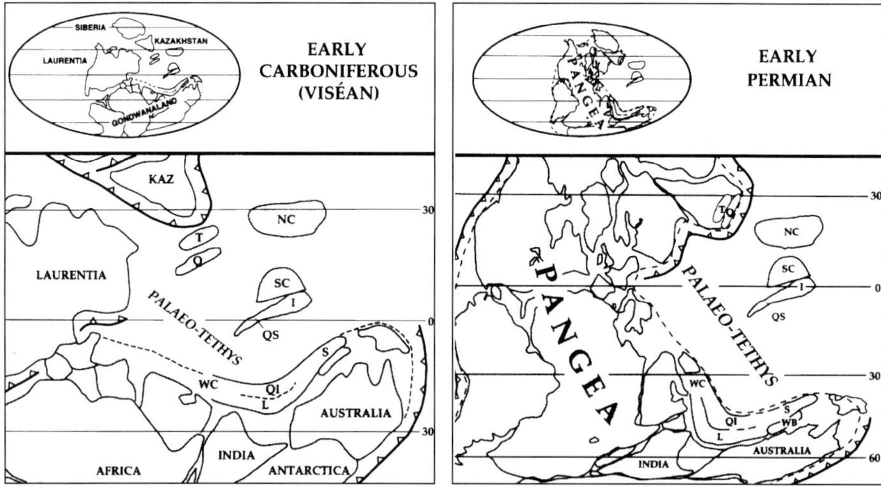

Figure 3.7. Maps of the Early Carboniferous and Early Permian showing the first rifting phase (Metcalfe, 1996).

Abbreviations in Fig. 3.7–3.9: Ba: Banda allochton; Ba-Su: Banggai-Sula; B-S: Buru-Seram; BU: Buton; C: Cimmerian continent; ES: E Sulawesi; I: Indochina; L: Lhasa; M: Mangkalihat; N: Natal; NC: N China; O: Obi-Bacan; QS: Qamdo-Simao; QT: Qiangtang; S: Sibumasu; SC: S China; SG: Songpan-Ganzi; Si: Sikuleh; Sm: Sumba; SWB: SW Borneo; T: Tarim; WB: W Burma; WC: W Cimmerian; WIJ: W Irian Jaya; WS: W Sulawesi;

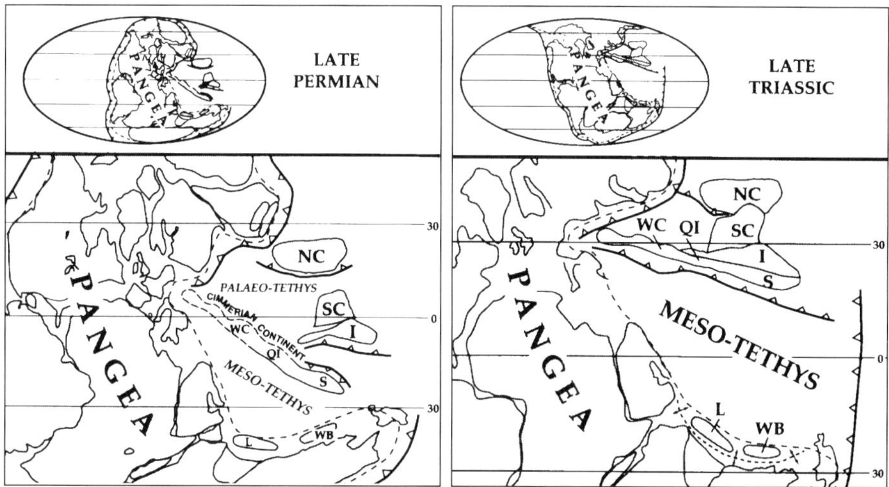

Figure 3.8. Maps of the Late Permian and Late Triassic showing the second rifting phase (Metcalfe, 1996).

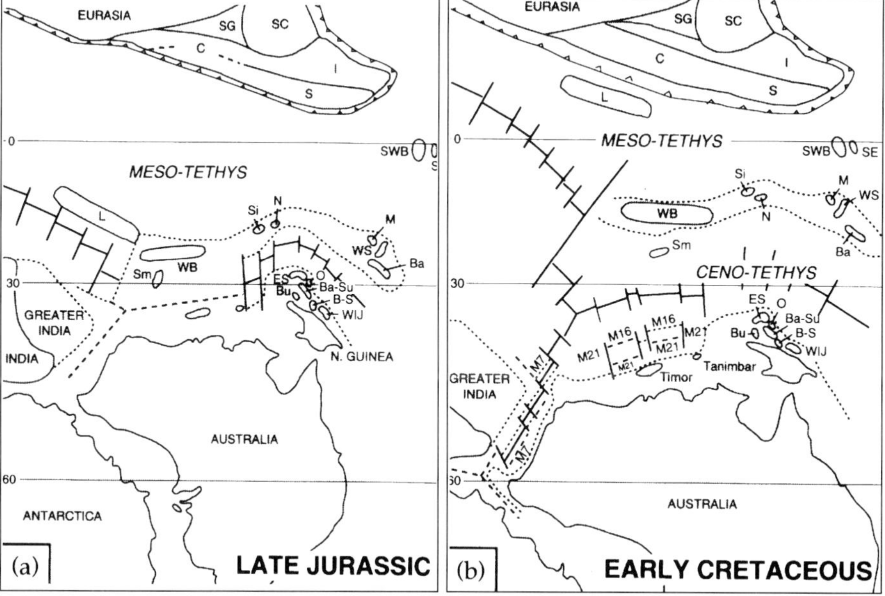

Figure 3.9. Maps of the Late Jurassic (a) and Early Cretaceous (b) showing the third rifting phase (Metcalfe, 1996).

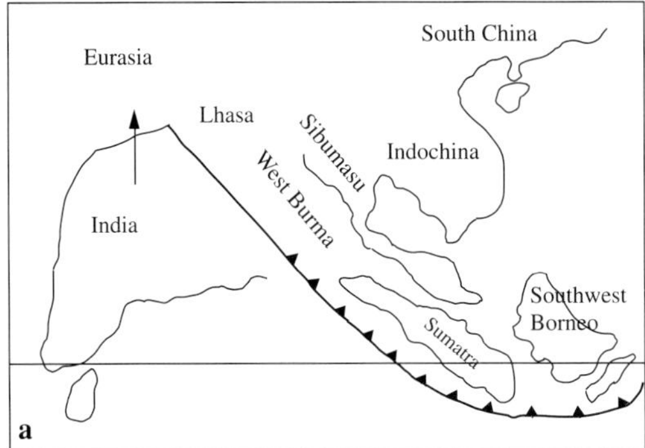

Figure 3.10. – a: Beginning of collision of India with Eurasia. – b: Some of the geological events that happened after the indentation of India into Eurasia. Timing is not necessarily the same for all events. Note that the West Burma Plate has moved northward, several basins have opened, and some parts of Southeast Asia have rotated clockwise or counterclockwise.

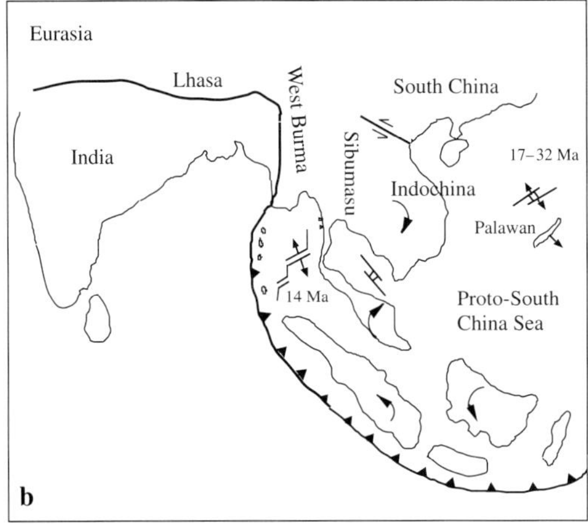

India initially also separated during the Late Jurassic (160 Ma). The onset of its collision (50 Ma) was probably the major cause for the clockwise rotation of Indochina, whereas the greater part of Sundaland performed a counterclockwise rotation (Fig. 3.10). Several basins and faults in SE Asia came into existence. This process was interrupted during the Middle Miocene by the collision of Australia with SE Asia (Rangin et al., 1990a, 1990b).

Other opinions on the major rifting events are of Audley-Charles (1987; et al., 1988) and Burrett et al. (1991). The latter placed the first rifting event of N and S China, Indochina and E Malaya in the Cambrian, the rifting of Sibumasu (also called Shan-Tai or Sinoburmalaya) in the Permian, and the last more continuous rifting phase in Cretaceous-Jurassic times. Audley-Charles (1987; et al., 1988), on the other hand, suggests that rifting occurred in later times: Burma, Thailand and Malaya, Sumatra and S Tibet separated from Gondwanaland in Early Jurassic times and collided with Asia at about 160 Ma. These land masses probably were exposed during most of the Jurassic and Cretaceous. The Late Cretaceous major eustatic sea level rise inundated large parts of the continents. However, Australia and the already rifted fragments remained largely above sea level.

Each collision event was accompanied by the closure of an existing (Palaeo-)Tethys Ocean, while each rifting event was followed by the opening of a new (Neo-)Tethys Ocean (Metcalfe, 1994b, 1996; Audley-Charles, 1987). The Uttaradit-Nan and Raub-Bentong line (Fig. 3.5), along which Indochina/East Malaya and Sinoburmalaya were sutured during the Late Permian/Early Triassic, can be regarded as the position of the former Palaeo-Tethys Ocean. To the north of the Himalayas, the Lancangjiang suture between the Qamdo-Simao and Qiangtang terranes (Fig. 3.5) represents another remnant of the closed Palaeo-Tethys. The Song Ma line (Fig. 3.5) is the suture along which Indochina / E Malaya welded with S China, together called Cathaysia (see Metcalfe) in the Early Carboniferous, before their collision with Eurasia. Cathaysia and N China were at that time located at low northern to equatorial latitudes.

A Permian Cathaysian flora of equatorial climate developed on blocks that were rifted during the first Palaezoic rifting event. Other blocks that remained close to Gondwanaland were influenced in Carbo-Permian times by cooler climatic factors, such as glaciation. Blocks that remained longer under Gondwanaland influence are: Lhasa (Tibet), Himalayas (part of Indian Plate), Shillong massif and Indian platform, Burma Plate, Sinoburmalaya and Shan-Tai (Sibumasu). Blocks of Cathaysian affinity are: Yangzi Platform (S China), Phu Hoat microcontinent (between S China and Indochina), Indochina and E Malaya, which is a continuation of Indochina beneath the South China Sea, and the West Borneo Basement (the terranes mentioned above in the first rifting phase). Sumatra probably has a mixed affinity: the Djambi province in S Sumatra with a Cathaysian type flora is probably a displaced part of the eastern Malay Peninsula, while northern Sumatra shows affinities with the western Malay Peninsula. Metcalfe regards northern Sumatra as part of Sibumasu, and southern Sumatra as belonging to E Malaya (Fig. 3.11).

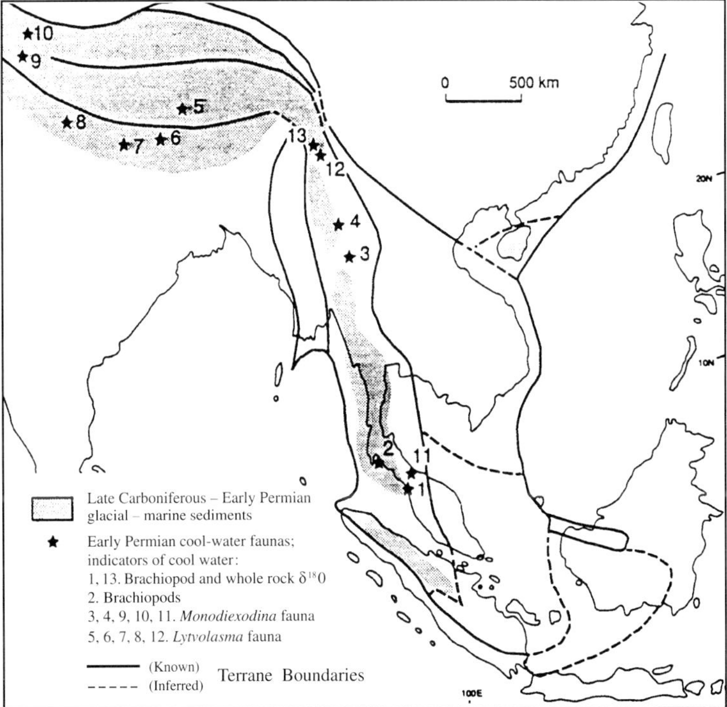

Figure 3.11. Map showing the distribution of cool-water indicators and glacial marine sediments in Southeast Asia and the Tibet-Himalaya region (Metcalfe, 1994a).

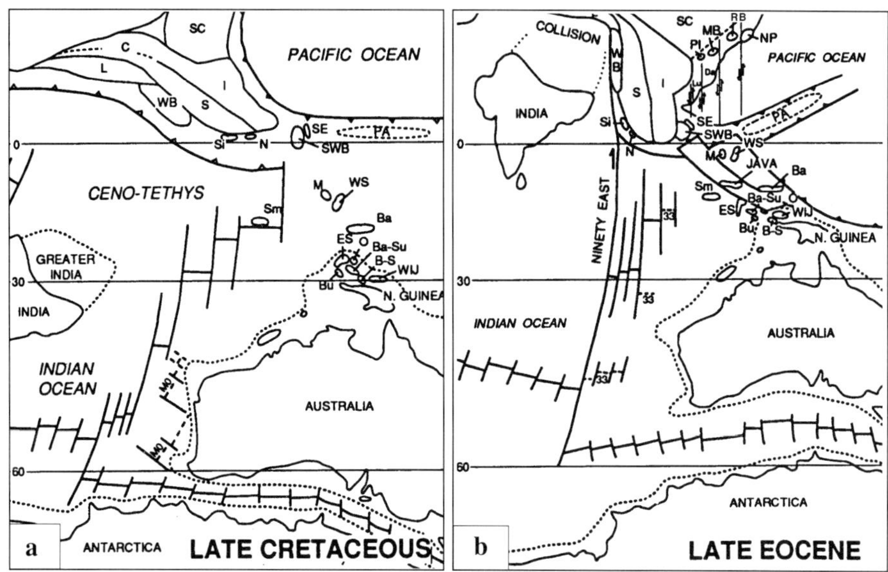

Figure 3.12. Maps of the Late Cretaceous (a) and Late Eocene (b) showing the rifting of India and Australia, and the collision of India (Metcalfe, 1994a). — For abbreviations, see page 79.

India

The initial separation of India from Gondwanaland (between the western margin of the Australian craton and the eastern margin of Africa) began in the Jurassic (Fig. 3.9). India moved northwest until 90 Ma due to sea floor spreading in the Argo Abyssal Plain (the oceanic part of the Indo-Australian Plate). Evidence from magnetic anomalies in the ocean floor shows a change in direction of the spreading just south of the Argo Abyssal Plain.

After 90 Ma, plate margins were reorganized, and Australia began to separate from Antarctica (see also Veevers, 1991). India rapidly moved (10 cm / year) with a changed direction northwards until 53 Ma (Early Eocene) when it started colliding with the Eurasian Plate (Fig. 3.12).

Weijermars (1989) in his reconstructions places India close to Africa during the Upper Cretaceous, whereas others isolate India during its rift towards Eurasia (see for discussion Briggs, 1989; Patterson & Owen, 1991; Thewissen & McKenna, 1992). These different views have implications concerning the development of a flora in India during its northward rift.

Until the collision of India with Eurasia, the Tethyan ocean (in fact Tethys III or Ceno-Tethys) subducted below Tibet; it was closed with the arrival of India. The Indus Yarlung Zangbo suture is a remnant of the closed Tethys (Fig. 3.5). At 45 Ma a major plate reorganization took place, and spreading in the Indian Ocean switched from northeast to southeast (Hutchison, 1992).

It is possible that spreading continued east of the Ninety East Ridge until 32 Ma, after which the Indian Plate and the Australian Plate were welded into one plate. The

Ninety East Ridge (Fig. 3.12) may be interpreted as the trace of the east margin of the Indian Plate; the Ridge has been built up of basalts, as the plate moved northwards over the Kerguelen hotspot.

The collision of India with the Eurasian Plate took part in two stages, a first one in the Early Eocene and a stronger one in the Late Eocene (45 Ma). The last one resulted in several major tectonic events such as the uplift of the Himalayas, the movements along pre-existing faults (Red River Fault), the clockwise rotation of Indochina and the subsequent opening of basins as the Gulf of Thailand as well as the Andaman Sea (Fig. 3.10). A large part of Sundaland, including the Malay Peninsula, Borneo and W Sulawesi, performed a counterclockwise rotation.

Burma

Before collision of the Indian continent eastward subduction occurred along the margin of the Asian Plate beneath the Burman Volcanic Arc, which became extinct after the collision of India with the Eurasian Plate (Fig. 3.5). The Indo-Burman Ranges, which represent an accretionary wedge formed during subduction of the Indian Plate below the Asian Plate, was uplifted during the Early Eocene as a result of the collision between India and Eurasia.

The West Burman Plate (also called Mount Victorialand) lay about 450 km south of its present location (Fig. 3.13). At this moment the small plate lies east to the Indo-Burman Ranges and west of the Sagaing-Namyin Fault. In the Early Mesozoic, the West Burman Plate probably formed an island arc which was separated by oceanic crust from the Asian Plate. The oceanic basin closed in the Late Cretaceous, and the island arc collided with Asia (Shan States, Sibumasu). Metcalfe (1994b) concludes that, although the origin of this block is not known, it might as well be part of a continental sliver that rifted off Gondwanaland in the Late Jurassic (WB in Fig. 3.9). During the Late Cenozoic the plate moved 450 km northwards along the Sagaing Fault. At that time the Andaman Sea was not yet formed.

The opening of the Andaman Sea started in the Middle Miocene (13 Ma) as the result of the oblique subduction of the Indian Plate.

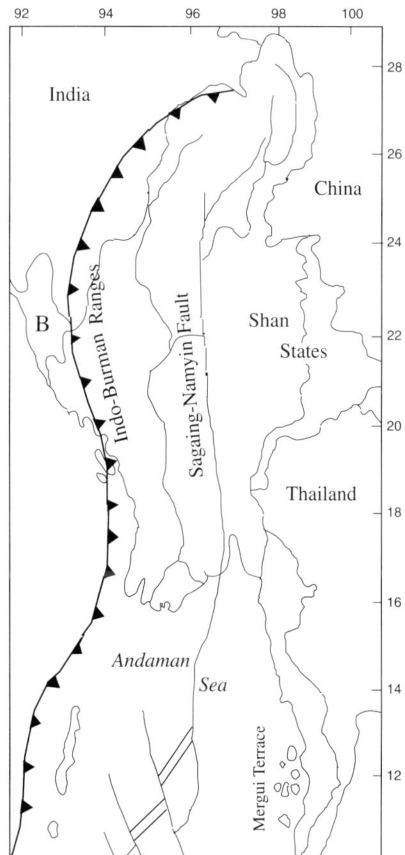

Figure 3.13. Map of Burma (redrawn after Hutchison, 1992). B = Bangladesh.

Before opening, the Andaman Sea probably represented a fore-arc basin in connection with the Andaman–Nicobar arc, the outer-arc ridge below which the Indian Plate is subducting. The Mergui Terrace (Fig. 3.13), adjacent to the Andaman Sea, represents a shelf extension of the Thai–Malay Peninsula block and is of continental crust. Some authors suggest that the Mergui area was a small continental fragment itself; Metcalfe feels there is little basis for this (references in Metcalfe, 1996). McCourt et al. (1996) indicate an extension of this microplate into W Sumatra.

Sinoburmalaya

Sinoburmalaya (or Sibumasu) started to separate from northern Gondwanaland together with other parts of the Cimmerian continent about 300 Ma (Fig. 3.8). Sinoburmalaya moved towards the Southeast Asian continent (at that moment consisting of Cathaysia), closing the Palaeo-Tethys Ocean that was located in between. The collision occurred during either the Late-Triassic or the Late Permian / Early Triassic and resulted in a fold belt that extends from the Malay Peninsula into Thailand and Yunnan along the Uttaradit-Nan and the Bentong-Raub Sutures (Fig. 3.5).

In the Jurassic the Eurasian Plate was above sea level. Afterwards it subsided resulting in the deposition of marine sediments in the Late Mesozoic. The plate was bordered in the south by a convergent margin (a volcanic arc) that extended from Sumatra through the Java Sea into the Meratus on Borneo.

Since the Late Eocene a clockwise rotation of at least 10° of Indochina and the Malay Peninsula occurred as the result of the collision of India with the Eurasian Plate. This same event is held responsible for the opening of the South China Sea and the Thai and Malay Basins.

Other scientists, however, regard the subduction of the Proto-South China Sea below Borneo as the major cause of opening of the South China Sea (Hall, 1995).

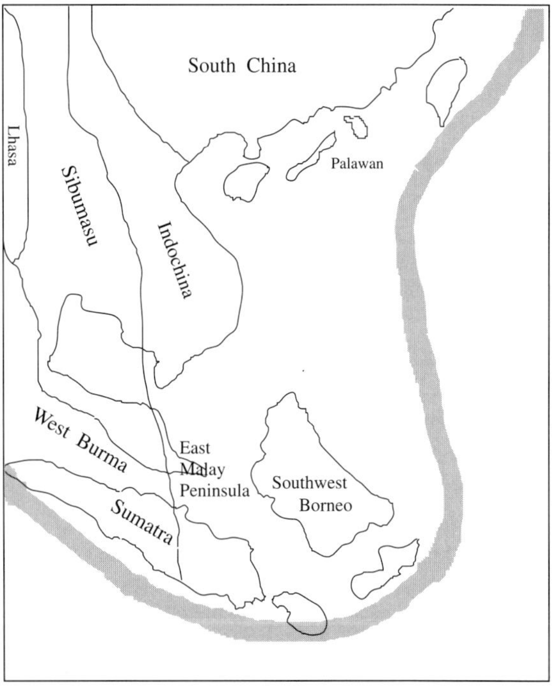

Figure 3.14. Simplified map of the ancient volcanic arc along the margin of the Asian Plate during the Cretaceous. The small West Burma Plate is shown to the east of Sibumasu.

The South China Sea region

A volcanic arc developed from the Mid-Jurassic to Mid-Cretaceous along the eastern margin of China. The arc extended from Hainan to Hong Kong and terminated in the east of Indochina (Fig. 3.14). Subduction occurred along this margin, but ceased before the onset of the Tertiary, after which the margin started to break up.

During the Jurassic the West Borneo Basement (Schwaner Block) separated from the Southeast Asian continent during the opening of the Proto-South China Sea. In the reconstructions of Metcalfe, however, SW Borneo and the Semitau remain at an equatorial latitude from the Late Jurassic onwards (Fig. 3.9).

The collision of India with Eurasia resulted in pericollisional extension in Southeast Asia, which led to the opening of the South China Sea (Fig. 3.15). As mentioned in the section on Sinoburmalaya, the opening of the South China Sea can also be explained by the pull forces of the subducting slab of the Proto-South China Sea below Borneo and the Philippine Sea Plate that resulted in stretching of the Eurasian margin (Hall, 1995, 1996; Hall et al., 1995). The South China Sea showed spreading activity from the Palaeocene or from the Mid-Oligocene to Early Miocene (32–17 Ma). The Proto-South China Sea subducted below Borneo and S Chinese continental fragments moved towards the island, which at that time performed an anticlockwise rotation of up to 50°.

Rifted fragments (Fig. 3.5) that are partly located above sea level, are represented by the Spratley Islands, Dangerous Grounds and the Reed Bank, and North Palawan and Mindoro (the Calamian group). The latter fragments form the oldest rocks of the Philippines. Other fragments are found on Borneo. Barber and Hall (1995) stressed the fact that, although in literature there is reference to fragments or blocks rifted off the South China margin, it is in fact one block that rifted southwards.

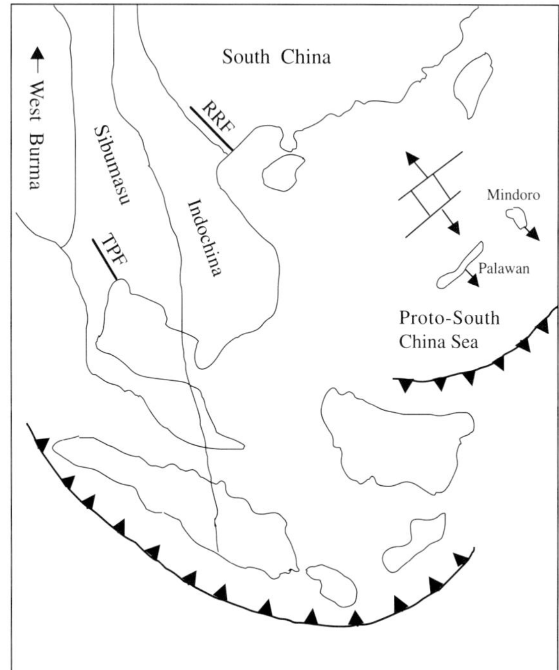

Figure 3.15. Schematic map of the opening of the South China Sea starting 17 Ma. Fragments of the Chinese continental margin move towards Borneo. Of the fragments only Palawan and Mindoro are shown. RRF= Red River Fault, TPF = Three Pagodas Fault.

The Proto-South China Sea was mainly consumed in the Palawan Trough according to Lee and Lawver (1992). The collision of the North Palawan microcontinent with the Cagayan ridge (West Philippines Block) in the Early Miocene stopped the opening of the South China Sea (Hall, 1996). After this collision subduction switched to the Sulu Sea, but was halted by the collision between the Sulu Ridge and Mindanao in the Late Miocene.

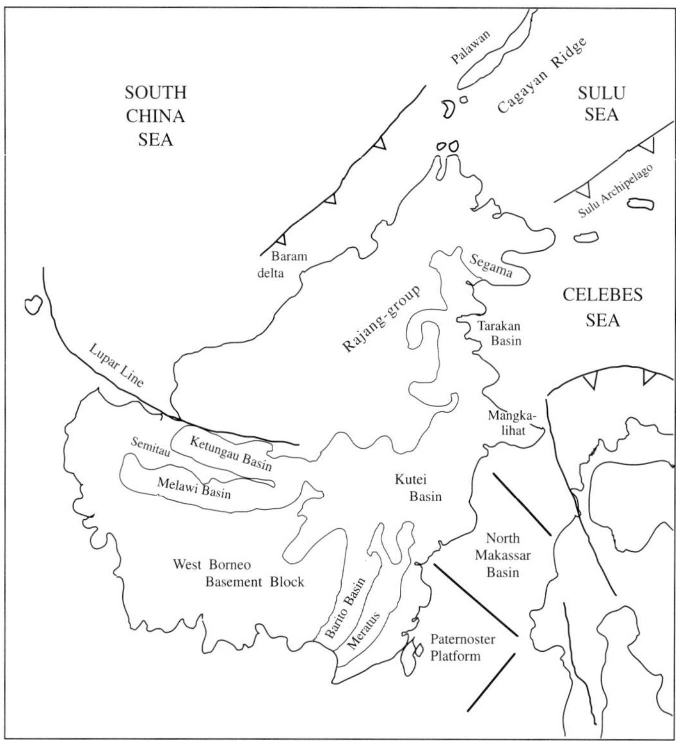

Figure 3.16. Map of Borneo showing major faults and basins.

Borneo

The island of Borneo is represented by a number of cratonic terranes that are separated by faults and younger orogenes, which stabilised during the Late Triassic. The Paternoster (Transform) Fault borders the Tertiary Kutei Basin towards the southern part of Borneo (Fig. 3.16). The West Borneo Basement Block is separated by the Boyan Suture from the Semitau terrane, which in turn is separated by the Lupar Line from the Northwest Borneo Geosyncline (Sarawak). The Boyan Suture is of Cretaceous age and represents the collision of the Semitau terrane and the (South) West Borneo Basement.

The West Borneo Basement is regarded by Hutchison as a displaced part of Indochina, which started rifting due to the opening of the Proto-South China Sea. Hamilton and Metcalfe placed this part of Borneo (SW Borneo and Semitau) in the Late Jurassic at equatorial latitude. Rangin et al. (1990b) suggest that the volcanic arc of the Meratus range (and E Java and Sumatra) was once continuous with the South China magmatic belt (Fig. 3.14). In their opinion Sundaland as a whole was displaced during the Late Cretaceous/Early Palaeocene along the Lupar line, which, in their view, is a continuation of the Three Pagodas Fault. Lorenz et al. (1993) accept a displacement of Indochina and Borneo of 700 km along the Red River.

In the Late Cretaceous continental fragments of the Chinese margin started to rift (see also Taylor & Hayes, 1983). As mentioned in the former paragraph, these fragments were rifting towards Borneo beneath which the Proto-South China Sea subducted. Partly incorporated in Borneo are (Fig. 3.5): The Luconia-Balingian block, for the largest part submerged off the coast of Borneo and reaching up to the Tinjar Fault, the Kelabit Highlands-Long Bawan province, and the Segama block near Tawau.

The Mangkalihat Peninsula block, lying between the Kutei and the Tarakan Basin (Fig. 3.16), is of different origin. According to Metcalfe it is part of the continental fragments (among them W Sulawesi) which rifted off the Gondwanaland margin in the Late Jurassic (M in Fig. 3.9), and contains the oldest fossils known from Borneo (Early Devonian corals).

In Palaeocene times (Hutchison, 1992) all of Borneo south and southeast of the Rajang Group was an extensive landmass together with what is now West Sulawesi and the area below the Java Sea.

The Kutei Basin was formed after the Early Tertiary opening of the Makassar Basin which rifted away the eastern part of Borneo to become what is now the west part of Sulawesi. The Melawi and Ketungai (Mandai) basins on the continental landmass of the West Borneo Basement are from the same period. In the northwest, the Rajang Basin (the now uplifted parts of Sarawak and Sabah) was a northwards directed gulf of the Proto-South China Sea in the Palaeocene and was uplifted probably in the Early Miocene. The Crocker Range represents the accretionary complex, and deltas (Baram, Tarakan) developed which received sediments from the uplifted parts of Borneo and probably from the Indochina part of the Sundashelf as well.

In Late Oligocene times a marine transgression phase is represented by the shallow marine carbonate platform south and east of the Rajang Basin.

Up to the Oligocene the Meratus Mountains in the southeast were not uplifted, but continuous with the basins at both sides. Only in the Late Miocene-Pliocene a rapid uplift of the Meratus Mountains took place.

At the same time the West Borneo Basement Block, dominated by the Schwaner Mountains, rotated counterclockwise about 50°.

To the northeast two island arcs bridge the gap between Borneo and the Philippines. The northwest one consists of Palawan and some other smaller islands, and belongs to the part of the southeast Chinese margin rifted off the Asian continent. To the southeast of these islands is the Sulu Sea. This sea is floored in the northern part by continental crust, to the south of the Cagayan ridge with ocean floor (Hall, 1995). Part of the Sulu

Sea subducted between 15 and 10 Ma in the south under the Sulu Archipelago. This forms the other island arc bridging Borneo and the Philippines, a now almost extinct volcanic arc.

Java

Java consists mainly of Cenozoic volcanic rocks with a marine basement. Continental crust is located in a broad belt in the Java Sea extending towards the Meratus Mountains in SE Borneo. East of this line there are no pre-Tertiary rocks. The belt may be regarded as a relict subduction zone that bounded the continent along its margin.

South of Java the Indian Plate is converging nearly perpendicular to the Java Trench. This trench is more than 6 km deep. The outer arc ridge is completely submerged, and the fore-arc basin is well developed.

From the Late Oligocene onwards, Java became part of the presently active arc/trench system that extends from the Indoburman Ranges to Java. The islands located to the east of Java (including East Java) are of Neogene age and are built on top of oceanic crust. The Sunda arc becomes younger towards the east, and continues as the Banda Arc.

Sumatra

The continental crust of Sumatra is built up of rocks of Lower Palaeozoic age. The volcanic arc extends parallel to the right lateral Semangko Fault system that can be regarded as the continuation of the West Andaman and the Sagaing-Namyin Faults. These faults originated due to the strong oblique subduction of the Indian Plate.

Northern Sumatra belonged to the Sinoburmalaya block, which collided with Indochina in the Late Triassic or Late Permian/Early Triassic (Fig. 3.8). Southern Sumatra, however, formed part of East Malaya and was welded to the Southeast Asian craton in the Early Carboniferous (Fig. 3.7).

Metcalfe describes two small terranes, Sikuleh and Natal (Woyla; Fig. 3.5), that rifted during the Late Jurassic off the Australian margin. These two terranes collided with North Sumatra in the Late Cretaceous.

To the west of Sumatra the accretionary wedge is visible by a chain of small islands on top of the ridge. This ridge continues towards the north as the Andaman–Nicobar Ridge.

In the Cenozoic Sumatra underwent a slight clockwise rotation.

The Philippines

The Philippines (Fig. 3.17) consist of a collage of blocks of diverse origin, amalgamated between two active arc-trench subduction systems. These are, respectively, the westward facing Manila trench in the west (subduction of South China Sea Basin) and the eastward facing Philippines Trench and East Luzon Trough system in the east (subduction of the West Philippines Sea Basin).

The Philippines do not consist of a single plate, but are the product of a diffuse deformation zone between the Philippine and the Eurasian Plate. Most of the islands

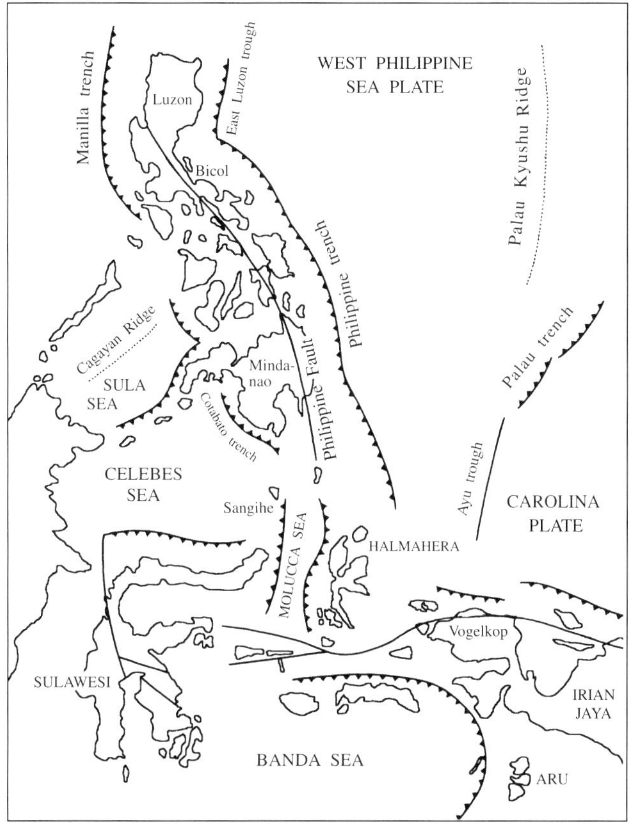

Figure 3.17. Geological map of the Philippine and Molucca region.

originally formed part of an arc at the southern edge of the Philippine Sea Plate. The Philippine Sea Plate has a Palaeocene age, and was formed at a spreading ridge at the southern boundary of the Pacific Plate. The Pacific Plate is moving to the northwest and subducting below Eurasia. Daly et al. (1991) suggest that spreading started 42 Ma ago, when the Pacific Plate motion changed from NNW to WNW. Spreading ceased at about 26 Ma. It is possible that the Central Basin Ridge is the fossil spreading centre of this Plate.

Since 35–40 Ma the Philippine Sea Basin has undergone a clockwise rotation of 50° and a northward drift. The Palau-Kyushu Ridge (volcanic arc) forms the eastern margin of the basin. The margin may continue into the Palau and Ayu Troughs which separate the Philippine Basin from the Carolina Plate (a small plate in the southwest of the large Pacific Plate). Subduction of the Carolina Plate beneath the Philippine Sea Basin has ceased or nearly so (Rangin et al., 1990a, 1990b). Hall (1995) placed the rotation of the plate before 40 Ma, after which rotation ceased and the spreading of the West Philippines Sea extended to the Celebes Sea and the Makassar Strait.

The eastern part of the Philippines (Luzon, Bicol, East and Central Mindanao) existed as a volcanic island arc which continues to Halmahera, and which is located more to the south along the margin of the West Philippine Sea Plate. The western Philippines (West Mindanao, Zambales) may be regarded as an extension of the volcanic arc that is located on the margin of the Eurasian Plate. Hall (1995) suggests that these western Philippine parts are from the Middle Eocene onwards (40 Ma) part of the Philippine Sea Plate, and that the Philippines are not the result of a collision of two opposite arcs.

In the Miocene the Philippines collided with the Eurasian margin. For a more extensive review of the Philippines see Hall (1995, 1996), Hall et al. (1995), and Rangin et al. (1990a, 1990b).

The Philippine Sea Plate rotated after the Miocene, and there was subduction at the Manilla Trench. Luzon subsequently collided with Taiwan. At present subduction probably ceased in the Manila Trench due to the collision of Taiwan with China. At the same time subduction begins at the Philippine Trench at the end of the Miocene. Plate consumption now only occurs along the Philippine Trench.

Sulawesi

The island of Sulawesi represents a very complex structure that originated from different sources. It is surrounded by arc-trench systems and sea basins.

West Sulawesi rifted from the eastern margin of Borneo in the Late Eocene (Fig. 3.16). The Paternoster Platform, which is located in the southern part of Makassar Strait, represents the part of Borneo to which the western arm of Sulawesi was connected. Metcalfe suggests in his reconstructions that West Sulawesi rifted from the Gondwanaland margin in Late Jurassic times. Masse et al. (1993) and Butterlin et al. (1993) describe the Paternoster Platform as a continental fragment (Argo) from the Australian Plate, which collided with Borneo in the Eocene. There is, however, no good evidence for this (Wensink, personal comm.). New insights on the evolution of the Makassar Strait are presented by Bergman et al. (1996).

West Sulawesi may be divided into two different terranes: a northern volcanic arc, and a western part that rifted from Borneo. Prior to the opening of Makassar Strait, the old Sulawesi subduction complex was assumed to be continuous with that of the Java Sea and Java (Fig. 3.14). The part that rifted from Borneo is of a continental nature and, before rifting, was volcanic during the period when it formed part of the subduction complex. This boundary is now obscured by the collision of ophiolitic masses and associated melanges, and some Australian continental fragments, e.g., Banggai-Sula and Buton (Wensink, 1996). The north arm of Sulawesi was part of the margin beneath which the northwards moving Indian–Australian Plate was subducting until the Early Miocene. From the Middle Miocene to the Quaternary, volcanism resumed and was accompanied by folding, faulting and uplift of the region. After spreading in the Celebes Sea, the arm began to rotate clockwise. At present the oceanic crust of the Celebes Sea is subducting southwards beneath Sulawesi's north arm.

The northeastern part of the Celebes Sea is subducting into the Cotabato Trench (below SW Mindanao). The Sangihe volcanic arc is located to the east of this basin, and continues into the northern arm of Sulawesi. This east facing arc has been active since Early or Middle Miocene. The Molluca Sea represents a separated part of the Indian Ocean lithosphere that was trapped by the northwards movement of the Australian continent. It is subducting both to the west under the Sangihe arc and to the east under the Halmahera arc, which is part of the Philippine Sea Plate. In fact the Molluca Sea Plate itself is probably almost completely subducted by the Philippine Sea Plate, which itself is subducting under the Eurasian Plate (Hall, 1995; Barber & Hall, 1995).

East Sulawesi contains suits of ophiolites of probably Indian Ocean origin. Buton and the Banggai-Sula spur located to the southeast and east of Sulawesi are continental fragments of the Australian Plate that were translated westwards from the Bird's Head of New Guinea along the Sorong Fault System. Burrett et al. (1991) show that Buton is a detached part of the Australian Plate that collided with the Sulawesi Trench 18 Ma ago. After this collision extension of the Banda arc began. The Banggai-Sula spur collided with the Sulawesi Trench in the Middle Miocene (10 Ma). The island arc of Sulawesi was bent by those collisions. Banggai-Sula subducted beneath Sulawesi's east arm.

The Australian margin

Separation of Australia from Antarctica occurred 53 Ma ago. The Australian Plate subducted and collided with the Banda arc trench system since the Middle-Late Miocene. At present the Australian Plate is underlying Timor, Sumba and the microcontinent of Buru-Seram, at the same time enclosing the Banda Sea from three sides. The heavy oceanic slab has been detached from the Australian continent below Timor, thus causing (by the diminished weight) a strong uplift of these islands (Burrett et al., 1991; Van der Werff, 1995).

New Guinea is located on the northern margin of Australia, and has accreted several terranes of Pacific origin. For a more extensive review, see Pigram & Davies (1987), Daly et al. (1991), and Rangin et al. (1990a, 1990b). An overview of the geological history of Australia is presented by Veevers (1991).

Chapter 4

HISTORICAL BIOGEOGRAPHY

INTRODUCTION

In this chapter the cladistic reconstruction of *Spatholobus* and allies, together with the cladograms of three other genera, will be used to construct a general area cladogram that gives information on the overall pattern of speciation (at least for these genera) in Southeast Asia, and in this way reflects the vicariance events and the history of the areas involved. In addition, an attempt has been made to use the cladogram of *Spatholobus* together with the general area cladogram to show the specific history of speciation and distribution of this genus within SE Asia.

Considerable research has been undertaken to find the most appropriate methods for analysis of data provided by phylogenies of taxa and their distribution. Several methods and criticisms on these methods have been given, in, e.g., Bremer (1992, 1995), Brooks (1990), Brooks & McLennan (1991), Cracraft (1983, 1988), Morrone & Carpenter (1994), Nelson & Ladiges (1991), Page (1988, 1990), Turner (1995a, 1995b), Van Welzen (1992), Wiley (1981, 1987), and Zandee & Roos (1987). Several computer programs are available to analyse taxon-area cladograms; most of these are in use for phylogenetic analyses as well. The best known of these are PAUP (Swofford, 1991, 1993, 1996) and Hennig86 (Farris, 1988) for BPA (Brooks Parsimony Analysis), CAFCA (Zandee, 1995) for CCA (Component Compatibility Analysis), COMPONENT (Page, 1993) for Component Analysis, and TAS (Nelson & Ladiges, 1992) for Three-Area Statements. Despite all efforts, however, all these methods and programs retain more or less serious flaws and "it must be admitted that current computer implementations of the methods remain unsatisfactory" (Morrone & Carpenter, 1994).

Apart from the flaws in the computer implementations, there are limitations in choice of the groups that can be used for the historical biogeographical analysis. In order to avoid comparing cladograms that reflect different time spans, it is best to use groups of organisms that have developed in the same time. Such groups may be expected to show the same vicariance events. Another important factor in the choice of the groups is the similar distribution pattern, especially in the Malesian area, where it is possible that plants have migrated from SE Asia into the archipelago eastward, or from the east towards SE Asia. If comparing opposite patterns, it is possible that the analysis will eliminate both patterns ('wave extinction') and reflect nothing. With these limitations in mind I used the results of the cladistic analysis of *Spatholobus* and allies, and those of the other groups, and compared these with the geological information as presented in Chapter 3. For practical reasons (computer equipment and time) I had to restrict myself to BPA and CCA analyses.

DATA

The data used for the biogeographical analysis were the phylogenies of *Spatholobus* and allies (Chapter 2), of *Fordia* (Schot, 1991), of *Genianthus* (Klackenberg, 1995), and of *Xanthophytum* (Axelius, 1990). The phylogenies of all these genera were generated using morphological characters, and parsimony analysis was undertaken with either Hennig86 or PAUP. The cladograms of the different genera are shown in Figure 4.1. The genera occur in the same region as *Spatholobus*: continental Southeast Asia and the West Malesian Archipelago (Fig. 4.2). Only *Xanthophytum* has species in New Guinea. This small monophyletic part in the cladogram was removed during the analysis to facilitate comparison with the other genera. The distribution data for each species per genus are given in Table 4.1.

METHODS

The first step was the delimitation of the areas used in the biogeographical analysis. For this step the distribution of all species of the genera *Spatholobus, Butea, Meizotropis, Xanthophytum, Fordia*, and *Genianthus* were taken into consideration. Although the genera occur more or less in the same region, there are differences in, e.g., the range of distribution (Fig. 4.2). There are widespread species and also species with very restricted areas within the genera. To facilitate comparison of the different area cladograms, areas were recognised based on the distribution of the species within each of the genera. These are not areas of endemism, but areas of distribution as described by Axelius (1991). Some areas contain endemics (= only occurring in the chosen area), other areas are composed of overlapping distributions of non-endemic species. Some authors, e.g. Sosef (1994), state that each area should at least contain one endemic, and that these 'remnant areas' (without endemics) are not informative. In my opinion, however, this is not necessary, and it is at least not feasible and practical to use only those areas which contain endemics. The species occurring in an area – endemic or not – will reflect their relation to other species in the coding of the areas, and thus a solution for placement of these 'remnant areas' is also possible. As a result of using these areas, it is possible that some end-branches will be empty, because the species coding for the area that is represented by the end-branch, is optimised on a node lower in the cladogram. On the other hand, difficulties may also arise with areas occupied by an endemic species having relations with species in different areas.

It is not always easy to decide on the delimitation of the areas, as it depends on the number of species present in an area. When there are too few species in an area (e.g., Sri Lanka) problems may arise during analysis. These areas often turn up basally in the cladogram due to lack of data (many zeroes in the area-taxon matrix), especially when an all zero outgroup is postulated, as is the case in BPA analysis. Furthermore, the delimitation of areas is related to their size. When the size chosen is too large (e.g., continental SE Asia) the results will remain uninformative. It is clear that by adding other genera to the analysis, the delimitation of the areas has to be reviewed and changed where necessary (e.g., more precise distribution areas within Sulawesi or East India/ Burma and Indochina). Figure 4.3 shows the basic geographic entities delimited for this study.

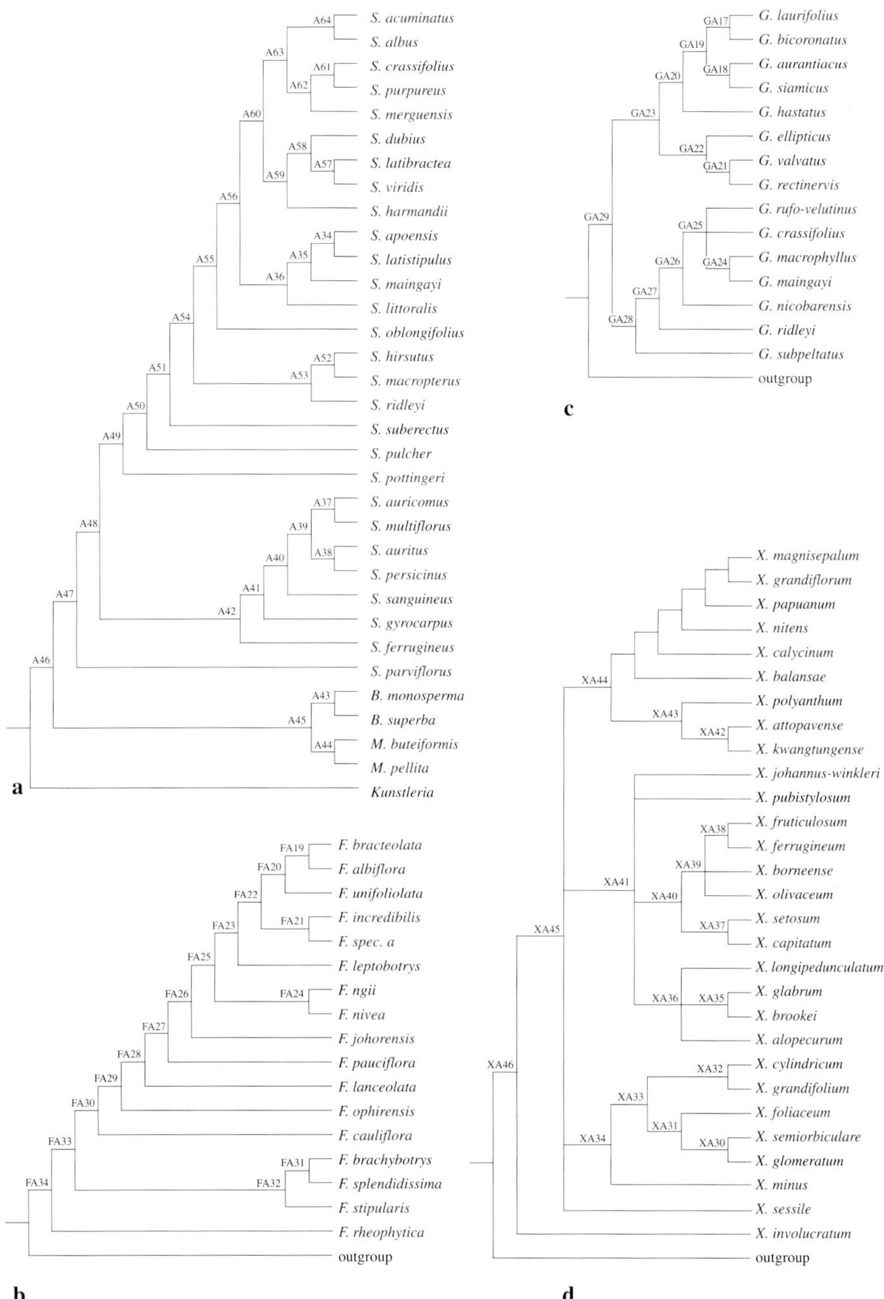

Figure 4.1. Cladograms of **a**: *Butea, Meizotropis*, and *Spatholobus*, **b**: *Fordia* (Schot, 1991), **c**: *Genianthus* (Klackenberg, 1995), **d**: *Xanthophytum* (Axelius, 1990). Numbers in the cladograms refer to the numbering of the ancestors used for compiling the area-taxon datamatrix.

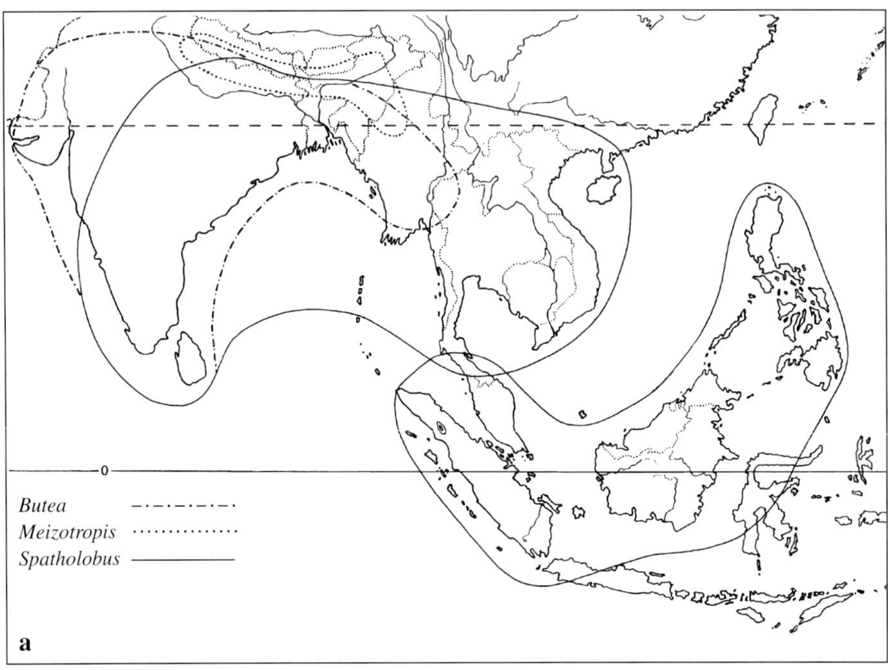

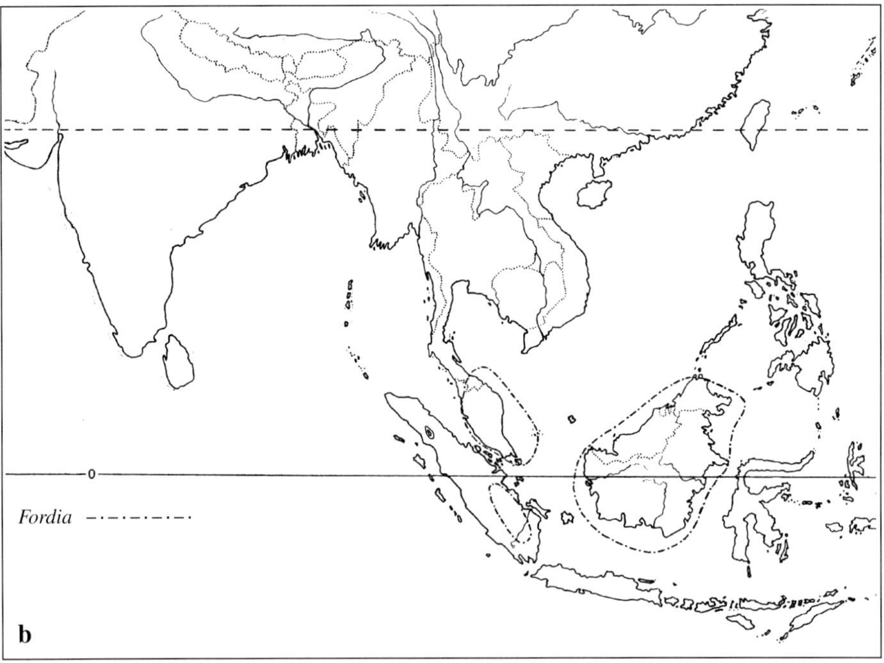

Figure 4.2. Distribution maps of the genera *Butea*, *Meizotropis*, *Spatholobus* (**a**), and *Fordia* (**b**).

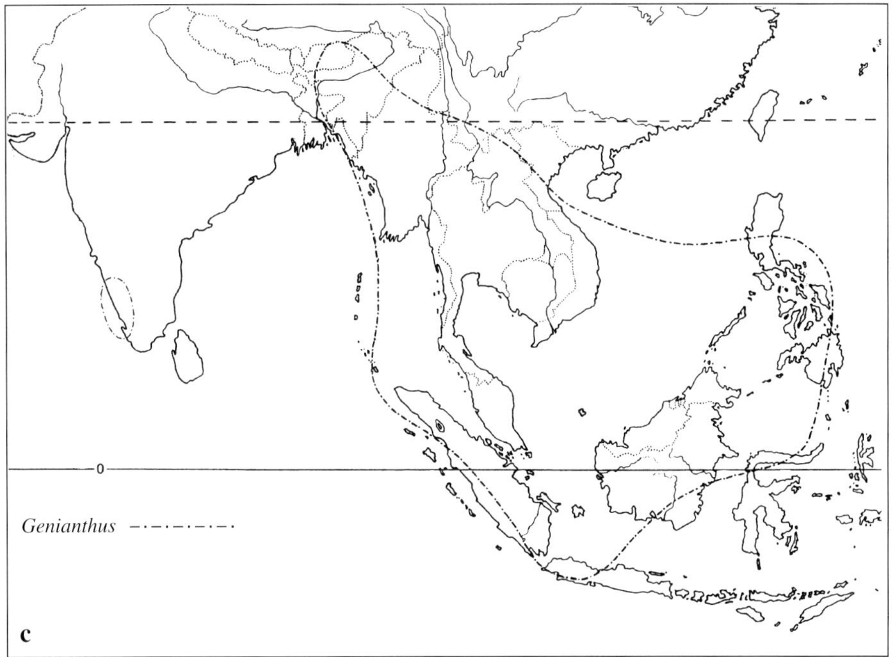

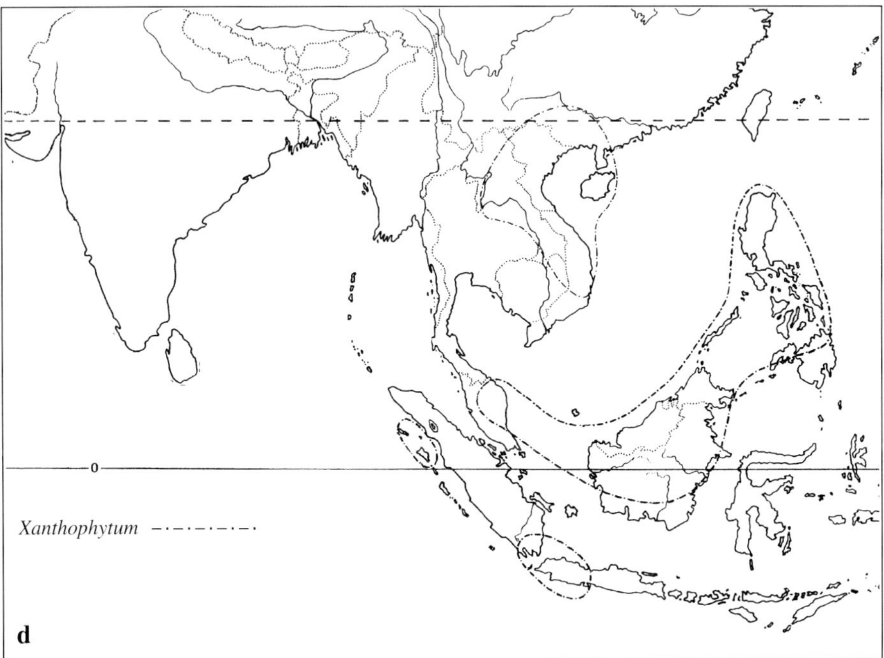

Figure 4.2. Distribution maps of the genera *Genianthus* (**c**) and *Xanthophytum* (**d**).

Table 4.1. Distribution of the species of *Butea, Fordia, Genianthus, Kunstleria, Meizotropis, Spatholobus,* and *Xanthophytum.*

Species	Areas of distribution	Species	Areas of distribution
Butea monosperma	1, 2, 3, 4, 5	*Spatholobus latibractea*	21, 22
Butea superba	2, 4, 5	*Spatholobus latistipulus*	19, 23, 24
Fordia albiflora	12, 13	*Spatholobus littoralis*	15, 18, 25, 27
Fordia brachybotrys	19, 20, 21, 22, 23, 24, 25, 28	*Spatholobus macropterus*	12,13,15,19,20,21, 22, 23, 24
Fordia bracteolata	12	*Spatholobus maingayi*	8, 12, 13, 14, 19, 20, 22, 23, 24, 27, 28
Fordia cauliflora	6		
Fordia incredibilis	14	*Spatholobus merguensis*	7
Fordia johorensis	14	*Spatholobus multiflorus*	21, 22, 23, 24
Fordia lanceolata	12	*Spatholobus oblongifolius*	21, 22, 23, 24
Fordia leptobotrys	5	*Spatholobus parviflorus*	1, 2, 3, 4, 5, 6
Fordia ngii	13	*Spatholobus persicinus*	19,20,21,22,24, 25
Fordia nivea	15	*Spatholobus pottingeri*	4, 5
Fordia ophirensis	12	*Spatholobus pulcher*	5
Fordia pauciflora	8, 11	*Spatholobus purpureus*	1
Fordia rheophyta	22, 23	*Spatholobus ridleyi*	14, 15
Fordia spec. a	14	*Spatholobus sanguineus*	22, 23, 24, 27
Fordia splendidissima	19,20,21,22,23,24,25	*Spatholobus suberectus*	5
Fordia stipularis	15	*Spatholobus viridis*	20, 21, 22, 23, 24
Fordia unifoliolata	12	*Xanthophytum alopecurum*	21
Genianthus aurantiacus	4, 5	*Xanthophytum attopavense*	5, 6
Genianthus bicoronatus	5	*Xanthophytum balansae*	6
Genianthus crassifolius	4, 6, 12, 15, 18, 22, 23	*Xanthophytum borneense*	21, 22, 24
Genianthus ellipticus	18, 21, 24, 27, 29	*Xanthophytum brookei*	21, 22
Genianthus hastatus	5	*Xanthophytum calycinum*	New Hebrides
Genianthus laurifolius	1, 6	*Xanthophytum capitatum*	19, 22, 23, 25
Genianthus macrophyllus	18	*Xanthophytum cylindricum*	21
Genianthus maingayi	12, 14, 20	*Xanthophytum ferrugineum*	13, 17, 26, 27, 28
Genianthus nicobarensis	10	*Xanthophytum foliaceum*	24
Genianthus rectinervis	27, 28	*Xanthophytum fruticulosum*	15, 18
Genianthus ridleyi	12	*Xanthophytum glabrum*	21, 22
Genianthus rufo-velutinus	12, 13	*Xanthophytum glomeratum*	21
Genianthus siamicus	6	*Xanthophytum grandiflorum*	New Guinea
Genianthus subpeltatus	12	*Xanthophytum grandifolium*	21
Genianthus valvatus	24	*Xanthophytum involucratum*	20
Kunstleria	1, 11–15, 20–28	*Xanthophytum johannis-winkleri*	19
Meizotropis buteiformis	2, 3, 4		
Meizotropis pellita	3	*Xanthophytum kwantungense*	5, 6
Spatholobus acuminatus	2, 4, 5, 6, 8, 9, 11	*Xanthophytum longipeduncu-latum*	23
Spatholobus albus	21, 22, 23		
Spatholobus apoensis	20, 28	*Xanthophytum magnisepalum*	New Guinea
Spatholobus auricomus	21, 21, 22	*Xanthophytum minus*	21
Spatholobus auritus	11, 19	*Xanthophytum nitens*	New Guinea
Spatholobus bracteolatus	7, 12	*Xanthophytum olivaceum*	19, 20
Spatholobus crassifolius	2, 4, 6, 11	*Xanthophytum papuanum*	New Guinea
Spatholobus dubius	11, 12	*Xanthophytum polyanthum*	6
Spatholobus ferrugineus	8, 11–26, 29	*Xanthophytum pubistylosum*	19, 20, 24
Spatholobus gyrocarpus	5, 8, 11, 12, 13, 16	*Xanthophytum semiorbiculare*	19, 21, 22, 24
Spatholobus harmandii	5, 6, 8	*Xanthophytum sessile*	20
Spatholobus hirsutus	19, 21, 22, 23, 24	*Xanthophytum setosum*	20

An area-taxon matrix (Table 4.2) was based on the available cladograms (Fig. 4.1) and the distributions (Table 4.1). The area-taxon matrix gives the presence (or absence) in a distribution area of each terminal node (= taxon) in the cladogram, as well as for all other (internal) nodes (= ancestors). In this way the structure of the cladogram is added to the matrix. The basic working hypothesis is that speciation results from passive allopatric speciation (vicariance), which implies no taxon-specific mechanisms. Therefore, it is assumed that the ancestor was present in the same areas as its descendants, thus excluding, e.g., dispersal events. The coding of the ancestor nodes is thus not independent, and is in fact an imperfect and unlikely assumption. In the case of incongruencies, dispersal is one of the possibilities to explain these patterns. Dispersal events may be recognised by parallelisms. This factor, often described as 'ad hoc' explanation, has to be kept in mind, because dispersal is an important strategy for plants, often leading to widespread species. In the case of extinction or primitive absence in part of the area, this way of coding leads to reversals (= species from state present to state absent). And, as mentioned above, if only one taxon (or a few) is present in an area (a column with many zeroes), the area may be placed artificially low in the area cladogram after analysis.

When one or more genera are absent in part of the total area, these missing data for areas (usually called 'missing areas') can be treated under different assumptions. Assumption 0 (Zandee & Roos, 1987) considers missing data as primitively absent. In this case missing data are coded as zero. Assumption 1 and 2 (Nelson & Platnick, 1981) treat them as uninformative. Under these assumptions missing data are coded as '?' (see in Table 4.2, e.g., Nicobar Islands for *Spatholobus*). In this analysis missing taxa were treated both under assumption 0 and assumption 1; widespread species were only treated under assumption 0, i.e., starting from the assumption that the areas inhabited by the widespread species formed one single area in the past (monophyletic origin). Under assumptions 1 and 2 the information given by the distribution of the widespread species is considered not as fully informative as under assumption 0 (mono- or paraphyletic origin, and in case of assumption 2 even polyphyletic).

The matrices used for each of the two methods differ slightly from each other (Table 4.2). BPA needs an artificial (all-zero) outgroup, and the area-taxon matrices of the genera have to be combined into one large matrix (30 areas, 170 taxa and ancestor taxa; as in Table 4.2). CCA needs no outgroup, but the matrices of each genus have to be entered separately for the analysis. In BPA the clades can be defined by the absence of a taxon (as is possible for characters), thus sometimes causing very artificial situations (Van Welzen, 1989, 1992). In CCA, however, the components used for constructing the cladogram are defined only by the presence of taxa. Sosef (1994) states that CCA may not find the correct cliques when dispersal or extinction events have occurred. In his view, it is better to use BPA in case of many dispersal and extinction events and check for clades solely based on the absence of taxa.

BPA was performed with PAUP 3.1.1 (Swofford, 1993). The complete matrix (Table 4.2) was analysed under heuristic search (addition sequence random, 1000×). The analysis was executed both under assumption 0 and 1 for missing areas.

Table 4.2. Taxon-area matrix of *Spatholobus*, *Butea*, *Meizotropis*, *Xanthophytum*, *Fordia*, and *Genianthus*. An all-zero outgroup has to be added for analysis with BPA. For CCA an outgroup is not added, and the datamatrix is divided into the parts belonging to the different genera. For analysis under assumption 0 (missing taxa = 0), the '?' has to be replaced by a zero; analysis under assumption 1 (genus is missing from area = ?) needs a '?'.

	S. acu	S. alb	S. apo	S. auric	S. aurit	S. cra	S. dub	S. fer	S. gyr	S. har	S. hir	S. latib	S. latis	S. lit	S. mac	S. mai	S. mer	S. mul	S. obl	S. par	S. per	S. pot	S. pul	S. pur	S. rid	S. san	S. sub	S. vir
1 Kerala	0	0	0	0	0	0	0	0	0	0	0	0	0	0	0	0	0	0	0	1	0	0	0	1	0	0	0	0
2 India	1	0	0	0	0	1	0	0	0	0	0	0	0	0	0	0	0	0	0	1	0	0	0	0	0	0	0	0
3 W Nepal	0	0	0	0	0	0	0	0	0	0	0	0	0	0	0	0	0	0	0	1	0	0	0	0	0	0	0	0
4 W Burma / E India	1	0	0	0	0	1	0	0	0	0	0	0	0	0	0	0	0	0	0	1	0	1	0	0	0	0	0	0
5 Yunnan / N Thailand	1	0	0	0	0	0	0	0	1	1	0	0	0	0	0	0	0	0	0	1	0	1	1	0	0	0	1	0
6 Indochina	1	0	0	0	0	1	0	0	0	1	0	0	0	0	0	0	0	0	0	1	0	0	0	0	0	0	0	0
7 Mergui / Tenasserim	0	0	0	0	0	0	0	0	0	0	0	0	0	0	0	0	1	0	0	0	0	0	0	0	0	0	0	0
8 S Thailand	1	0	0	0	0	0	0	1	1	1	0	0	0	0	0	1	0	0	0	0	0	0	0	0	0	0	0	0
9 Andaman Is	1	0	0	0	0	0	0	0	0	0	0	0	0	0	0	0	0	0	0	0	0	0	0	0	0	0	0	0
10 Nicobar Is	?	?	?	?	?	?	?	?	?	?	?	?	?	?	?	?	?	?	?	?	?	?	?	?	?	?	?	?
11 Penang / Kedah	1	0	0	0	1	1	1	1	1	1	0	0	0	0	0	0	0	0	0	0	0	0	0	0	0	0	0	0
12 W Malaya	0	0	0	0	0	0	1	1	1	0	0	0	0	0	0	1	1	0	0	0	0	0	0	0	0	0	0	0
13 E Malaya	0	0	0	0	0	0	0	1	1	0	0	0	0	0	0	1	1	0	0	0	0	0	0	0	0	0	0	0
14 Singapore	0	0	0	0	0	0	0	1	0	0	0	0	0	0	0	0	1	0	0	0	0	0	0	0	0	1	0	0
15 SE Sumatra	0	0	0	0	0	0	0	1	0	0	0	0	0	1	1	0	0	0	0	0	0	0	0	0	0	1	0	0
16 N Sumatra	0	0	0	0	0	0	0	1	1	0	0	0	0	0	0	0	0	0	0	0	0	0	0	0	0	0	0	0
17 Sumatra coast	0	0	0	0	0	0	0	1	0	0	0	0	0	0	0	0	0	0	0	0	0	0	0	0	0	0	0	0
18 W Java	0	0	0	0	0	0	0	1	0	0	0	0	0	1	0	0	0	0	0	0	0	0	0	0	0	0	0	0
19 SW Borneo	0	0	0	0	1	0	0	1	0	0	1	0	1	0	1	1	0	0	0	0	1	0	0	0	0	0	0	0
20 Semitau	0	0	1	1	0	0	0	1	0	0	0	0	0	0	1	1	0	0	0	0	1	0	0	0	0	0	0	1
21 Sarawak / C Borneo	0	1	0	1	0	0	0	1	0	0	1	1	0	0	1	0	0	1	1	0	1	0	0	0	0	0	0	1
22 NW Borneo	0	1	0	1	0	0	0	1	0	0	1	1	0	0	1	1	0	1	1	0	1	0	0	0	0	1	0	1
23 NE Borneo	0	1	0	0	0	0	0	1	0	0	1	0	1	0	1	1	0	1	1	0	0	0	0	0	0	1	0	1
24 CE Borneo	0	0	0	0	0	0	0	1	0	0	1	0	1	0	1	1	0	1	1	0	1	0	0	0	0	1	0	1
25 Meratus	0	0	0	0	0	0	0	1	0	0	0	0	0	1	0	0	0	0	0	0	1	0	0	0	0	0	0	0
26 Palawan	0	0	0	0	0	0	0	1	0	0	0	0	0	0	0	0	0	0	0	0	0	0	0	0	0	0	0	0
27 N Philippines	0	0	0	0	0	0	0	0	0	0	0	0	0	1	0	1	0	0	0	0	0	0	0	0	0	1	0	0
28 S Philippines	0	0	1	0	0	0	0	0	0	0	0	0	0	0	0	1	0	0	0	0	0	0	0	0	0	0	0	0
29 Sulawesi	0	0	0	0	0	0	0	1	0	0	0	0	0	0	0	0	0	0	0	0	0	0	0	0	0	0	0	0
30 Outgroup area	0	0	0	0	0	0	0	0	0	0	0	0	0	0	0	0	0	0	0	0	0	0	0	0	0	0	0	0

	B. mon	B. sup	M. but	M. pel	A34	A35	A36	A37	A38	A39	A40	A41	A42	A43	A44	A45	A46	A47	A48	A49	A50	A51	A52	A53	A54	A55	A56	A57
1 Kerala	1	0	0	0	0	0	0	0	0	0	0	0	0	1	0	1	1	1	1	1	1	1	0	0	1	1	1	0
2 India	1	1	1	0	0	0	0	0	0	0	0	0	0	1	1	1	1	1	1	1	1	1	0	0	1	1	1	0
3 W Nepal	1	0	1	1	0	0	0	0	0	0	0	0	0	1	1	1	1	1	0	0	0	0	0	0	0	0	0	0
4 W Burma / E India	1	1	1	0	0	0	0	0	0	0	0	0	0	1	1	1	1	1	1	1	1	1	0	0	1	1	1	0
5 Yunnan / N Thailand	1	1	0	0	0	0	0	0	0	0	0	1	1	1	0	1	1	1	1	1	1	1	0	0	1	1	1	0
6 Indochina	0	0	0	0	0	0	0	0	0	0	0	0	0	0	0	0	1	1	1	1	1	1	0	0	1	1	1	0
7 Mergui / Tenasserim	0	0	0	0	0	0	0	0	0	0	0	0	0	0	0	0	1	1	1	1	1	1	0	0	1	1	1	0
8 S Thailand	0	0	0	0	0	1	1	0	0	0	0	1	1	0	0	0	1	1	1	1	1	1	0	0	1	1	1	0
9 Andaman Is	0	0	0	0	0	0	0	0	0	0	0	0	0	0	0	0	1	1	1	1	1	1	0	0	1	1	1	0
10 Nicobar Is	?	?	?	?	?	?	?	?	?	?	?	?	?	?	?	?	?	?	?	?	?	?	?	?	?	?	?	?
11 Penang / Kedah	0	0	0	0	0	0	0	0	1	1	1	1	1	0	0	0	1	1	1	1	1	1	0	0	1	1	1	0
12 W Malaya	0	0	0	0	0	1	1	0	0	0	0	1	1	0	0	0	1	1	1	1	1	1	1	1	1	1	1	0
13 E Malaya	0	0	0	0	0	1	1	0	0	0	0	1	1	0	0	0	1	1	1	1	1	1	1	1	1	1	1	0
14 Singapore	0	0	0	0	0	1	1	0	0	0	0	0	0	1	0	0	0	1	1	1	1	1	0	1	1	1	1	0
15 SE Sumatra	0	0	0	0	0	0	1	0	0	0	0	0	1	0	0	0	1	1	1	1	1	1	1	1	1	1	1	0
16 N Sumatra	0	0	0	0	0	0	0	0	0	0	0	0	1	1	0	0	0	1	1	1	0	0	0	0	0	0	0	0
17 Sumatra coast	0	0	0	0	0	0	0	0	0	0	0	0	0	0	0	0	0	1	1	1	0	0	0	0	0	0	0	0
18 W Java	0	0	0	0	0	0	1	0	0	0	0	0	1	0	0	0	1	1	1	1	1	1	0	0	1	1	1	0
19 SW Borneo	0	0	0	0	1	1	1	0	1	1	1	1	1	0	0	0	1	1	1	1	1	1	1	1	1	1	1	0
20 Semitau	0	0	0	0	1	1	1	1	1	1	1	1	1	0	0	0	1	1	1	1	1	1	1	1	1	1	1	1
21 Sarawak / C Borneo	0	0	0	0	0	0	0	1	1	1	1	1	1	0	0	0	1	1	1	1	1	1	1	1	1	1	1	1
22 NW Borneo	0	0	0	0	0	1	1	1	1	1	1	1	1	0	0	0	1	1	1	1	1	1	1	1	1	1	1	1
23 NE Borneo	0	0	0	0	1	1	1	1	0	1	1	1	1	0	0	0	1	1	1	1	1	1	1	1	1	1	1	1
24 CE Borneo	0	0	0	0	1	1	1	1	1	1	1	1	1	0	0	0	1	1	1	1	1	1	1	1	1	1	1	1
25 Meratus	0	0	0	0	0	0	1	0	1	1	1	1	1	0	0	0	1	1	1	1	1	1	0	0	1	1	1	0
26 Palawan	0	0	0	0	0	0	0	0	0	0	0	0	1	0	0	0	1	1	1	0	0	0	0	0	0	0	0	0
27 N Philippines	0	0	0	0	0	1	1	0	0	0	1	1	1	0	0	0	1	1	1	1	1	0	0	0	1	1	1	0
28 S Philippines	0	0	0	0	1	1	1	0	0	0	0	0	0	0	0	0	1	1	1	1	1	1	0	0	1	1	1	0
29 Sulawesi	0	0	0	0	0	0	0	0	0	0	0	0	1	0	0	0	1	1	1	0	0	0	0	0	0	0	0	0
30 Outgroup area	0	0	0	0	0	0	0	0	0	0	0	0	0	0	0	0	0	0	0	0	0	0	0	0	0	0	0	0

(Table 4.2 continued)

	A58	A59	A60	A61	A62	A63	A64	X. alo	X. att	X. bal	X. bor	X. bro	X. cal	X. cap	X. cyl	X. fer	X. fol	X. fru	X. gla	X. gra. fl	X. gra. fo	X. glo	X. inv	X. joh	X. kwa	X. lon	X. mag	X. min
1 Kerala	0	0	1	1	1	1	0	?	?	?	?	?	?	?	?	?	?	?	?	?	?	?	?	?	?	?	?	?
2 India	0	0	1	1	1	1	1	?	?	?	?	?	?	?	?	?	?	?	?	?	?	?	?	?	?	?	?	?
3 W Nepal	0	0	0	0	0	0	0	?	?	?	?	?	?	?	?	?	?	?	?	?	?	?	?	?	?	?	?	?
4 W Burma / E India	0	0	1	1	1	1	1	?	?	?	?	?	?	?	?	?	?	?	?	?	?	?	?	?	?	?	?	?
5 Yunnan / N Thailand	0	1	1	0	0	1	1	0	1	0	0	0	?	0	0	0	0	0	0	?	0	0	0	0	1	0	?	0
6 Indochina	0	1	1	1	1	1	1	0	1	1	0	0	?	0	0	0	0	0	0	?	0	0	0	0	1	0	?	0
7 Mergui / Tenasserim	0	0	1	0	1	1	0	?	?	?	?	?	?	?	?	?	?	?	?	?	?	?	?	?	?	?	?	?
8 S Thailand	0	1	1	0	0	1	1	?	?	?	?	?	?	?	?	?	?	?	?	?	?	?	?	?	?	?	?	?
9 Andaman Is	0	0	1	0	0	1	1	?	?	?	?	?	?	?	?	?	?	?	?	?	?	?	?	?	?	?	?	?
10 Nicobar Is	?	?	?	?	?	?	?	?	?	?	?	?	?	?	?	?	?	?	?	?	?	?	?	?	?	?	?	?
11 Penang / Kedah	1	1	1	1	1	1	1	?	?	?	?	?	?	?	?	?	?	?	?	?	?	?	?	?	?	?	?	?
12 W Malaya	1	1	1	0	0	0	0	?	?	?	?	?	?	?	?	?	?	?	?	?	?	?	?	?	?	?	?	?
13 E Malaya	0	0	0	0	0	0	0	0	0	0	0	0	?	0	0	1	0	0	0	?	0	0	0	0	0	0	?	0
14 Singapore	0	0	0	0	0	0	0	?	?	?	?	?	?	?	?	?	?	?	?	?	?	?	?	?	?	?	?	?
15 SE Sumatra	0	0	0	0	0	0	0	0	0	0	0	0	?	0	0	0	0	1	0	?	0	0	0	0	0	0	?	0
16 N Sumatra	0	0	0	0	0	0	0	?	?	?	?	?	?	?	?	?	?	?	?	?	?	?	?	?	?	?	?	?
17 Sumatra coast	0	0	0	0	0	0	0	0	0	0	0	0	?	0	0	1	0	0	0	?	0	0	0	0	0	0	?	0
18 W Java	0	0	0	0	0	0	0	0	0	0	0	0	?	0	0	0	0	1	0	?	0	0	0	0	0	0	?	0
19 SW Borneo	0	0	0	0	0	0	0	0	0	0	0	0	?	1	0	0	0	0	0	?	0	0	0	1	0	0	?	0
20 Semitau	1	1	1	0	0	0	0	0	0	0	0	0	?	0	0	0	0	0	0	?	0	0	1	0	0	0	?	0
21 Sarawak / C Borneo	1	1	1	0	0	1	1	1	0	0	1	1	?	0	1	0	0	0	1	?	1	1	0	0	0	0	?	1
22 NW Borneo	1	1	1	0	0	1	1	0	0	0	1	1	?	1	0	0	0	0	1	?	0	0	0	0	0	0	?	0
23 NE Borneo	1	1	1	0	0	1	1	0	0	0	0	0	?	1	0	0	0	0	0	?	0	0	0	0	0	1	?	0
24 CE Borneo	1	1	1	0	0	0	0	0	0	0	1	0	?	0	0	0	1	0	0	?	0	0	0	0	0	0	?	0
25 Meratus	0	0	0	0	0	0	0	0	0	0	0	0	?	1	0	0	0	0	0	?	0	0	0	0	0	0	?	0
26 Palawan	0	0	0	0	0	0	0	0	0	0	0	0	?	0	0	1	0	0	0	?	0	0	0	0	0	0	?	0
27 N Philippines	0	0	0	0	0	0	0	0	0	0	0	0	?	0	0	1	0	0	0	?	0	0	0	0	0	0	?	0
28 S Philippines	0	0	0	0	0	0	0	0	0	0	0	0	?	0	0	1	0	0	0	?	0	0	0	0	0	0	?	0
29 Sulawesi	0	0	0	0	0	0	0	?	?	?	?	?	?	?	?	?	?	?	?	?	?	?	?	?	?	?	?	?
30 Outgroup area	0	0	0	0	0	0	0	0	0	0	0	0	?	0	0	0	0	0	0	?	0	0	0	0	0	0	?	0

	X. nit	X. oli	X. pap	X. pol	X. pub	X. sem	X. ses	X. set	XA30	XA31	XA32	XA33	XA34	XA35	XA36	XA37	XA38	XA39	XA40	XA41	XA42	XA43	XA44	XA45	XA46	F. alb	F. brach	F. bract
1 Kerala	?	?	?	?	?	?	?	?	?	?	?	?	?	?	?	?	?	?	?	?	?	?	?	?	?	?	?	?
2 India	?	?	?	?	?	?	?	?	?	?	?	?	?	?	?	?	?	?	?	?	?	?	?	?	?	?	?	?
3 W Nepal	?	?	?	?	?	?	?	?	?	?	?	?	?	?	?	?	?	?	?	?	?	?	?	?	?	?	?	?
4 W Burma / E India	?	?	?	?	?	?	?	?	?	?	?	?	?	?	?	?	?	?	?	?	?	?	?	?	?	?	?	?
5 Yunnan / N Thailand	?	0	?	0	0	0	0	0	0	0	0	0	0	0	0	0	0	0	0	0	1	1	1	1	1	0	0	0
6 Indochina	?	0	?	1	0	0	0	0	0	0	0	0	0	0	0	0	0	0	0	0	1	1	1	1	1	0	0	0
7 Mergui / Tenasserim	?	?	?	?	?	?	?	?	?	?	?	?	?	?	?	?	?	?	?	?	?	?	?	?	?	?	?	?
8 S Thailand	?	?	?	?	?	?	?	?	?	?	?	?	?	?	?	?	?	?	?	?	?	?	?	?	?	0	0	0
9 Andaman Is	?	?	?	?	?	?	?	?	?	?	?	?	?	?	?	?	?	?	?	?	?	?	?	?	?	?	?	?
10 Nicobar Is	?	?	?	?	?	?	?	?	?	?	?	?	?	?	?	?	?	?	?	?	?	?	?	?	?	?	?	?
11 Penang / Kedah	?	?	?	?	?	?	?	?	?	?	?	?	?	?	?	?	?	?	?	?	?	?	?	?	?	0	0	0
12 W Malaya	?	?	?	?	?	?	?	?	?	?	?	?	?	?	?	?	?	?	?	?	?	?	?	?	?	1	0	1
13 E Malaya	?	0	?	0	0	0	0	0	0	0	0	0	0	0	0	0	1	1	1	1	0	0	0	1	1	1	0	0
14 Singapore	?	?	?	?	?	?	?	?	?	?	?	?	?	?	?	?	?	?	?	?	?	?	?	?	?	0	0	0
15 SE Sumatra	?	0	?	0	0	0	0	0	0	0	0	0	0	0	0	0	1	1	1	1	0	0	0	1	1	0	0	0
16 N Sumatra	?	?	?	?	?	?	?	?	?	?	?	?	?	?	?	?	?	?	?	?	?	?	?	?	?	?	?	?
17 Sumatra coast	?	0	?	0	0	0	0	0	0	0	0	0	0	0	0	0	1	1	1	1	0	0	0	1	1	?	?	?
18 W Java	?	0	?	0	0	0	0	0	0	0	0	0	0	0	0	0	1	1	1	1	0	0	0	1	1	?	?	?
19 SW Borneo	?	1	?	0	1	1	0	0	1	1	0	1	1	0	0	1	0	1	1	1	0	0	0	1	1	0	1	0
20 Semitau	?	1	?	0	1	0	1	1	0	0	0	0	0	0	0	1	0	1	1	1	0	0	0	1	1	0	1	0
21 Sarawak / C Borneo	?	0	?	0	0	1	0	0	1	1	1	1	1	1	1	0	0	1	1	1	0	0	0	1	1	0	1	0
22 NW Borneo	?	0	?	0	0	1	0	0	1	1	0	1	1	1	1	1	0	1	1	1	0	0	0	1	1	0	1	0
23 NE Borneo	?	0	?	0	0	0	0	0	0	0	0	0	0	0	1	1	0	0	1	1	0	0	0	1	1	0	1	0
24 CE Borneo	?	0	?	0	1	1	0	0	1	1	0	1	1	0	0	0	0	1	1	1	0	0	0	1	1	0	1	0
25 Meratus	?	0	?	0	0	0	0	0	0	0	0	0	0	0	0	1	0	0	1	1	0	0	0	1	1	0	1	0
26 Palawan	?	0	?	0	0	0	0	0	0	0	0	0	0	0	0	0	1	1	1	1	0	0	0	1	1	?	?	?
27 N Philippines	?	0	?	0	0	0	0	0	0	0	0	0	0	0	0	0	1	1	1	1	0	0	0	1	1	?	?	?
28 S Philippines	?	0	?	0	0	0	0	0	0	0	0	0	0	0	0	0	1	1	1	1	0	0	0	1	1	0	1	0
29 Sulawesi	?	?	?	?	?	?	?	?	?	?	?	?	?	?	?	?	?	?	?	?	?	?	?	?	?	?	?	?
30 Outgroup area	?	0	?	0	0	0	0	0	0	0	0	0	0	0	0	0	0	0	0	0	0	0	0	0	0	0	0	0

(Table 4.2 continued)

	F. cau	F. inc	F. joh	F. lan	F. lep	F. ngii	F. niv	F. oph	F. pau	F. spl	F. rhe	F. sti	F. uni	F. spec.	FA19	FA20	FA21	FA22	FA23	FA24	FA25	FA26	FA27	FA28	FA29	FA30	FA31	FA32	FA33
1 Kerala	?	?	?	?	?	?	?	?	?	?	?	?	?	?	?	?	?	?	?	?	?	?	?	?	?	?	?	?	?
2 India	?	?	?	?	?	?	?	?	?	?	?	?	?	?	?	?	?	?	?	?	?	?	?	?	?	?	?	?	?
3 W Nepal	?	?	?	?	?	?	?	?	?	?	?	?	?	?	?	?	?	?	?	?	?	?	?	?	?	?	?	?	?
4 W Burma / E India	?	?	?	?	?	?	?	?	?	?	?	?	?	?	?	?	?	?	?	?	?	?	?	?	?	?	?	?	?
5 Yunnan/N Thailand	0	0	0	0	1	0	0	0	0	0	0	0	0	0	0	0	0	0	1	0	1	1	1	1	1	1	0	0	1
6 Indochina	1	0	0	0	0	0	0	0	0	0	0	0	0	0	0	0	0	0	0	0	0	0	0	0	0	1	0	0	1
7 Mergui/Tenasserim	?	?	?	?	?	?	?	?	?	?	?	?	?	?	?	?	?	?	?	?	?	?	?	?	?	?	?	?	?
8 S Thailand	0	0	0	0	0	0	0	0	1	0	0	0	0	0	0	0	0	0	0	0	0	0	1	1	1	1	0	0	1
9 Andaman Is	?	?	?	?	?	?	?	?	?	?	?	?	?	?	?	?	?	?	?	?	?	?	?	?	?	?	?	?	?
10 Nicobar Is	?	?	?	?	?	?	?	?	?	?	?	?	?	?	?	?	?	?	?	?	?	?	?	?	?	?	?	?	?
11 Penang / Kedah	0	0	0	0	0	0	0	0	1	0	0	0	0	0	0	0	0	0	0	0	0	0	1	1	1	1	0	0	1
12 W Malaya	0	0	0	1	0	0	0	1	0	0	0	0	1	0	1	1	0	1	1	0	1	1	1	1	1	1	0	0	1
13 E Malaya	0	0	0	0	0	1	0	0	0	0	0	0	0	0	0	1	0	1	1	1	1	1	1	1	1	1	0	0	1
14 Singapore	0	1	1	0	0	0	0	0	0	0	0	0	0	0	1	0	1	1	1	0	1	1	1	1	1	1	0	0	1
15 SE Sumatra	0	0	0	0	0	0	1	0	0	0	0	1	0	0	0	0	0	0	0	1	1	1	1	1	1	1	0	1	1
16 N Sumatra	?	?	?	?	?	?	?	?	?	?	?	?	?	?	?	?	?	?	?	?	?	?	?	?	?	?	?	?	?
17 Sumatra coast	?	?	?	?	?	?	?	?	?	?	?	?	?	?	?	?	?	?	?	?	?	?	?	?	?	?	?	?	?
18 W Java	?	?	?	?	?	?	?	?	?	?	?	?	?	?	?	?	?	?	?	?	?	?	?	?	?	?	?	?	?
19 SW Borneo	0	0	0	0	0	0	0	0	0	1	0	0	0	0	0	0	0	0	0	0	0	0	0	0	0	0	1	1	1
20 Semitau	0	0	0	0	0	0	0	0	0	1	0	0	0	0	0	0	0	0	0	0	0	0	0	0	0	0	1	1	1
21 Sarawak/C Borneo	0	0	0	0	0	0	0	0	0	1	0	0	0	0	0	0	0	0	0	0	0	0	0	0	0	0	1	1	1
22 NW Borneo	0	0	0	0	0	0	0	0	0	1	1	0	0	0	0	0	0	0	0	0	0	0	0	0	0	0	1	1	1
23 NE Borneo	0	0	0	0	0	0	0	0	0	1	1	0	0	0	0	0	0	0	0	0	0	0	0	0	0	0	1	1	1
24 CE Borneo	0	0	0	0	0	0	0	0	0	1	0	0	0	0	0	0	0	0	0	0	0	0	0	0	0	0	1	1	1
25 Meratus	0	0	0	0	0	0	0	0	0	1	0	0	0	0	0	0	0	0	0	0	0	0	0	0	0	0	1	1	1
26 Palawan	?	?	?	?	?	?	?	?	?	?	?	?	?	?	?	?	?	?	?	?	?	?	?	?	?	?	?	?	?
27 N Philippines	?	?	?	?	?	?	?	?	?	?	?	?	?	?	?	?	?	?	?	?	?	?	?	?	?	?	?	?	?
28 S Philippines	0	0	0	0	0	0	0	0	0	0	0	0	0	0	0	0	0	0	0	0	0	0	0	0	0	0	1	1	1
29 Sulawesi	?	?	?	?	?	?	?	?	?	?	?	?	?	?	?	?	?	?	?	?	?	?	?	?	?	?	?	?	?
30 Outgroup area	0	0	0	0	0	0	0	0	0	0	0	0	0	0	0	0	0	0	0	0	0	0	0	0	0	0	0	0	0

	FA34	G. lau	G. bic	G. aur	G. sia	G. has	G. ell	G. val	G. rec	G. mac	G. mai	G. cra	G. rufo	G. nic	G. rid	G. sub	GA17	GA18	GA19	GA20	GA21	GA22	GA23	GA24	GA25	GA26	GA27	GA28	GA29
1 Kerala	?	1	0	0	0	0	0	0	0	0	0	0	0	0	0	0	1	0	1	1	0	0	1	0	0	0	0	0	1
2 India	?	?	?	?	?	?	?	?	?	?	?	?	?	?	?	?	?	?	?	?	?	?	?	?	?	?	?	?	?
3 W Nepal	?	?	?	?	?	?	?	?	?	?	?	?	?	?	?	?	?	?	?	?	?	?	?	?	?	?	?	?	?
4 W Burma / E India	?	0	0	1	0	0	0	0	0	0	0	1	0	0	0	0	0	1	1	1	0	0	1	0	1	1	1	1	1
5 Yunnan/N Thailand	1	0	1	1	0	1	0	0	0	0	0	0	0	0	0	0	1	1	1	1	0	0	1	0	0	0	0	0	1
6 Indochina	1	1	1	0	0	1	0	0	0	0	0	0	1	0	0	0	0	1	1	1	1	0	0	1	0	1	1	1	1
7 Mergui/Tenasserim	?	?	?	?	?	?	?	?	?	?	?	?	?	?	?	?	?	?	?	?	?	?	?	?	?	?	?	?	?
8 S Thailand	1	?	?	?	?	?	?	?	?	?	?	?	?	?	?	?	?	?	?	?	?	?	?	?	?	?	?	?	?
9 Andaman Is	?	?	?	?	?	?	?	?	?	?	?	?	?	?	?	?	?	?	?	?	?	?	?	?	?	?	?	?	?
10 Nicobar Is	?	0	0	0	0	0	0	0	0	0	0	0	0	1	0	0	0	0	0	0	0	0	0	0	0	1	1	1	1
11 Penang / Kedah	1	?	?	?	?	?	?	?	?	?	?	?	?	?	?	?	?	?	?	?	?	?	?	?	?	?	?	?	?
12 W Malaya	1	0	0	0	0	0	0	0	0	0	1	1	1	0	1	1	0	0	0	0	0	0	0	1	1	1	1	1	1
13 E Malaya	1	0	0	0	0	0	0	0	0	0	0	0	1	0	0	0	0	0	0	0	0	0	0	0	1	1	1	1	1
14 Singapore	1	0	0	0	0	0	0	0	0	0	1	0	0	0	0	0	0	0	0	0	0	0	0	0	1	1	1	1	1
15 SE Sumatra	1	0	0	0	0	0	0	0	0	0	0	1	0	0	0	0	0	0	0	0	0	0	0	0	1	1	1	1	1
16 N Sumatra	?	?	?	?	?	?	?	?	?	?	?	?	?	?	?	?	?	?	?	?	?	?	?	?	?	?	?	?	?
17 Sumatra coast	?	?	?	?	?	?	?	?	?	?	?	?	?	?	?	?	?	?	?	?	?	?	?	?	?	?	?	?	?
18 W Java	?	0	0	0	0	0	1	0	0	1	0	1	0	0	0	0	0	0	0	0	0	1	1	1	1	1	1	1	1
19 SW Borneo	1	?	?	?	?	?	?	?	?	?	?	?	?	?	?	?	?	?	?	?	?	?	?	?	?	?	?	?	?
20 Semitau	1	0	0	0	0	0	0	0	0	0	1	0	0	0	0	0	0	0	0	0	0	0	0	0	1	1	1	1	1
21 Sarawak/C Borneo	1	0	0	0	0	0	1	0	0	0	0	0	0	0	0	0	0	0	0	0	0	1	1	0	0	0	0	0	1
22 NW Borneo	1	0	0	0	0	0	0	0	0	0	0	1	0	0	0	0	0	0	0	0	0	0	0	0	1	1	1	1	1
23 NE Borneo	1	0	0	0	0	0	0	0	0	0	0	1	0	0	0	0	0	0	0	0	0	0	0	0	1	1	1	1	1
24 CE Borneo	1	0	0	0	0	0	1	1	0	0	0	0	0	0	0	0	0	0	0	0	1	1	1	0	0	0	0	0	1
25 Meratus	1	?	?	?	?	?	?	?	?	?	?	?	?	?	?	?	?	?	?	?	?	?	?	?	?	?	?	?	?
26 Palawan	?	?	?	?	?	?	?	?	?	?	?	?	?	?	?	?	?	?	?	?	?	?	?	?	?	?	?	?	?
27 N Philippines	?	0	0	0	0	0	1	0	1	0	0	0	0	0	0	0	0	0	0	0	1	1	1	0	0	0	0	0	1
28 S Philippines	1	0	0	0	0	0	0	0	1	0	0	0	0	0	0	0	0	0	0	0	1	1	1	0	0	0	0	0	1
29 Sulawesi	?	0	0	0	0	0	1	0	0	0	0	0	0	0	0	0	0	0	0	0	0	1	1	0	0	0	0	0	1
30 Outgroup area	0	0	0	0	0	0	0	0	0	0	0	0	0	0	0	0	0	0	0	0	0	0	0	0	0	0	0	0	0

For CCA the computer program CAFCA was used (Zandee, 1995). The four matrices (for each group its own matrix) were analysed under the biogeographical run, generalised area cladogram, with as cladon option partial monothetic sets [defined by the unique occurrence of (at least one) monophyletic group of taxa], and as cladogram selection criterium the minimum amount of state changes (steps). Again the analysis was performed under both assumption 0 and 1.

AREAS

The distribution areas shown in Figure 4.3 are based only on the distribution of the species. It is remarkable that a large part of these areas coincide with the tectonic terranes as described by, e.g., Metcalfe and Hutchison (see Chapter 3 and Figure 3.5).

In the text below, geological (and other historical) information on each of the separate areas is summarised. This summary is based on the information on the geological history of Southeast Asia, as provided in Chapter 3. Where necessary, the abbreviation of the name of the area (as used in the computer analyses) is given in parentheses. These abbreviations (versus the full geographical names) are also used further on in the text of the present chapter. In addition to the historical information of the areas, the species of the genera *Butea, Fordia, Genianthus, Meizotropis, Spatholobus* and *Xanthophytum* occurring in the areas are mentioned in the summary below.

1. **Kerala** is part of the Indian continent. This part of India has retained tropical rain forest during all of the last 18,000 years. In other parts of India, there have been extensive grasslands during dry periods, but this seems to have been one of the rain forest refuges.
   In Kerala, *S. purpureus* is endemic; *B. monosperma, G. laurifolius,* and *S. parviflorus* also occur here.
2. **India** (including Kerala, which is taken as a separate area) is the continental part of the Indian Plate that separated as a large fragment from Australia in the Jurassic (c. 160 Ma). It started colliding with Eurasia in the Early Eocene (53 Ma), but the final collision took place in the Late Eocene (45 Ma). After this event the Himalayas were uplifted.
   In this large area both species of *Butea, M. buteiformis, S. acuminatus, S. crassifolius,* and *S. parviflorus* are found.
3. **West Nepal** and **Kumaon** (W Nepal) are part of the Himalayan region, which was uplifted after the Late Eocene. This area changed significantly after the uplift and provided a geographical barrier to the north.
   Meizotropis pellita is endemic in this region. The other species occurring here are *B. monosperma, M. buteiformis,* and *S. parviflorus.*
4. **Bangladesh** and **eastern India up to Burma** (W Burma / E India). From a geological point of view, this area is composed of several parts. Bangladesh forms part of the Indian Plate; W Burma and the mountains of eastern India are part of the West Burma Plate, which was uplifted in the Early Eocene by the subducting Indian Plate. Later, during the Cenozoic, the small West Burma Plate moved northwards to its present position by the movement of the Indian Plate.

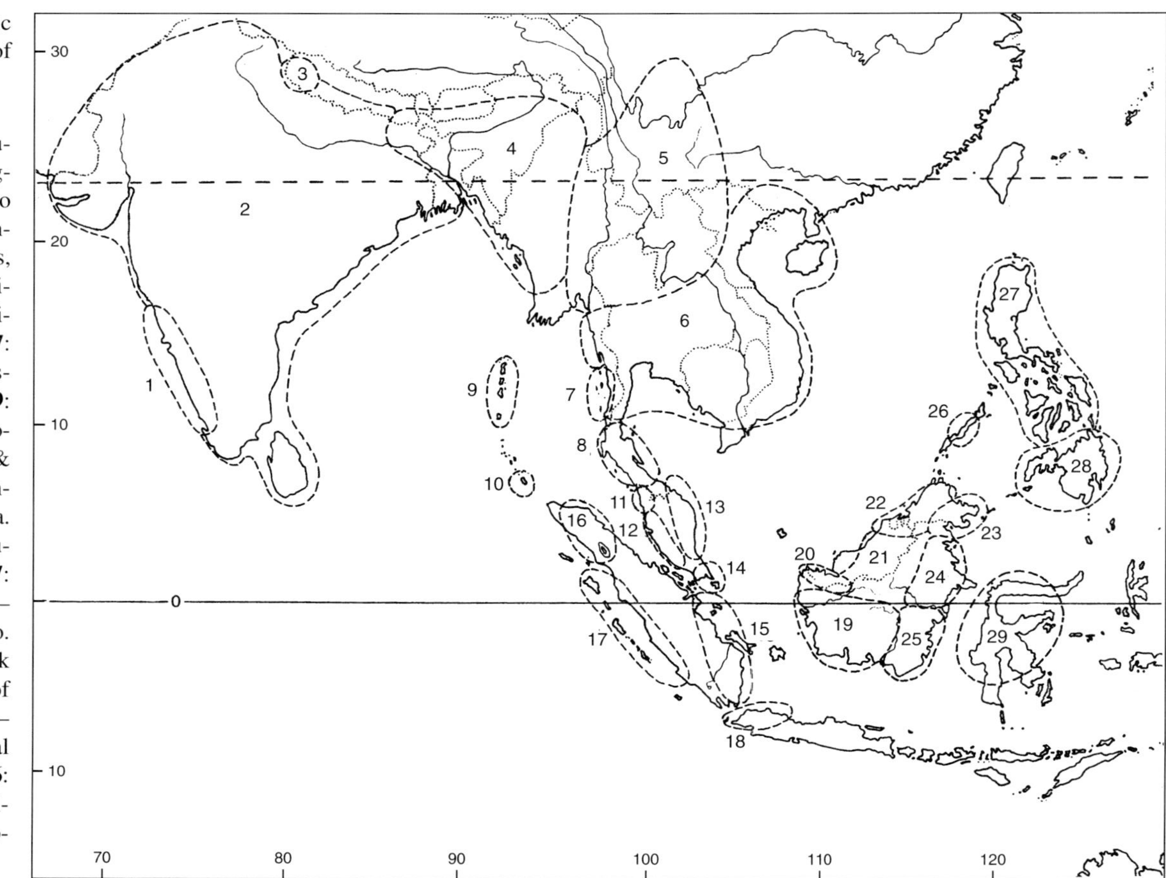

Figure 4.3. Map with basic geographic entities (areas of distribution).

1: Kerala. – **2**: India. – **3**: Kumaon & W Nepal. – **4**: Bangladesh & eastern India up to Burma. – **5**: E Burma, northern Thailand & northern Laos, Yunnan. – **6**: Indochina, Hainan, S Guangxi, central Thailand and part of Burma. – **7**: Mergui Archipelago/Tenasserim. – **8**: S Thailand. – **9**: Andaman Islands. – **10**: Nicobar Islands. – **11**: Penang & Kedah. – **12**: W Malay Peninsula. – **13**: E Malay Peninsula. – **14**: Singapore. – **15**: SE Sumatra. – **16**: N Sumatra. – **17**: Sumatra coast and islands. – **18**: W Java. – **19**: SW Borneo. – **20**: Semitau. – **21**: Sarawak and central mountain range of Borneo. – **22**: NW Borneo. – **23**: NE Borneo. – **24**: Central E Borneo. – **25**: Meratus. – **26**: Palawan. – **27**: Northern Philippines. – **28**: Southern Philippines. – **29**: Sulawesi.

Four species of *Spatholobus* occur (*S. acuminatus*, *S. crassifolius*, *S. parviflorus*, *S. pottingeri*), both *Butea* species, *G. aurantiacus*, *G. crassifolius*, and *M. buteiformis*.

5. ***East Burma, northern Thailand, northern Laos*** and ***Yunnan*** (Yunnan / N Thailand) are all part of different continental fragments. East Burma and the north of Thailand (west part) are part of Sibumasu; N Laos and the northeast of Thailand are part of Indochina; Yunnan is the southwest part of S China. South China and Indochina rifted away from Gondwanaland in the Palaeozoic, and sutured along the Song Ma line in the Early Carboniferous. These parts developed a Cathaysian flora subject to an equatorial climate. Parts that remained near the Gondwana margin in Carbo-Permian times were influenced by the cooler climate. Sibumasu rifted off the Gondwana margin in a later phase, during the Early to Middle Permian. Here glacial-marine elements in the fossil fauna indicate an influence due to the cooler climate. Indochina and Sibumasu collided in the Late Permian / Early Triassic. It is clear that these parts have already been welded into one area for a long period.

   Endemic in the area are *F. leptobotrys*, *G. bicoronatus*, *G. hastatus*, *S. pulcher*, and *S. suberectus*. Species occurring here are both species of *Butea*, *G. auranticus*, *S. gyrocarpus*, *S. harmandii*, *S. pottingeri*, *X. attopavense*, and *X. kwantungense*.

6. ***Indochina, Hainan, South Guangxi, Central Thailand*** and ***part of Burma*** (Indochina) are parts of Indochina, South China, and Sibumasu. This composite area has a history comparable to area 5.

   Endemic in this area are *F. cauliflora*, *G. siamicus*, *X. balansae*, and *X. polyanthum*. Other species in this area are *S. acuminatus*, *S. crassifolius*, *S. harmandii*, *S. parviflorus*, *G. crassifolius*, *G. laurifolius*, and *X. attopavense*.

7. The ***Mergui Archipelago*** and ***Tenasserim*** (Mergui / Tenasserim) are part of Sibumasu. The Mergui Archipelago is in fact an extension of the continental shelf, while Tenasserim is part of the continent itself.

   Here only *S. merguensis* is endemic. The other species occurring here, *S. bracteolatus*, is not included in the analysis due to incompleteness of the available material.

8. ***South Thailand*** (S Thailand) is the narrow part of the country, near the Isthmus of Kra. It mainly represents the transition between the more seasonal and everwet climates during the present. This part has been inundated extensively in the past, during periods of high sea level (Chapter 3: Fig. 3.3). Tropical rain forest has established here at least 18,000 years ago. At that time the sea level was low and a savannah corridor from Indochina to Java was present, bordered by monsoon forest (Chapter 3: Fig. 3.3). Only the largest parts of Sumatra and Malaya, extending up to the Isthmus of Kra, were carrying rain forest. Other tropical rain forest refugia were present in Borneo, the Philippines and Sulawesi.

   No endemics occur in this region. The species found here are *F. pauciflora* , *S. acuminatus*, *S. ferrugineus*, *S. gyrocarpus*, *S. harmandii*, and *S. maingayi*.

9 & 10. The ***Andaman Islands*** & ***Nicobar Islands*** represent the outer arc ridge, below which the Indian Plate subducted before the opening of the Andaman Sea in the Middle Miocene (13 Ma).

   On the Andaman Islands only the widespread *S. acuminatus* occurs; on the Nicobar Islands only the endemic species *G. nicobarensis* is present.

11, 12, & 16. ***Penang & Kedah*** (Penang / Kedah), the ***West Malay Peninsula*** (W Malaya), & ***North Sumatra*** (N Sumatra) are part of Sibumasu (see also area 5). In area 11 (Penang / Kedah) no endemics are present. *Spatholobus auritus* is probably endemic in area 11, if the identification of the sterile specimen of *S. auritus* in SW Borneo (area 19) is wrong. Other species occurring here are *F. pauciflora, S. acuminatus, S. crassifolius* (?), *S. dubius, S. ferrugineus*, and *S. gyrocarpus*. In area 12 (W Malaya) three species of *Fordia* are present as endemics: *F. lanceolata, F. ophirensis*, and *F. unifoliolata*. Other species occurring here are *F. albiflora, S. bracteolatus, S. dubius, S. ferrugineus, S. gyrocarpus, S. macropterus*, and *S. maingayi*. In area 16 (N Sumatra) *S. ferrugineus* and *S. gyrocarpus* are present.

13, 14, & 15. The ***East Malay Peninsula*** (E Malaya), ***Singapore***, & ***Southeast Sumatra*** (SE Sumatra) are part of the same fragment as Indochina. The Bentong-Raub line is the suture between East Malaya and West Malaya (Late Triassic). The whole part of East and West Malaya has been rotated clockwise after the collision of India.
In area 13 (E Malaya) *F. ngii* is endemic. Other species in this area are *F. albiflora, S. ferrugineus, S. gyrocarpus, S. macropterus, S. maingayi.*, and *X. ferrugineum*. In area 14 (Singapore) three species of *Fordia* are endemic: *F. incredibilis, F. johorensis,* and *F. spec. a*. Other species here are *G. maingayi, S. ferrugineus, S. maingayi*, and *S. ridleyi*. In area 15 (SE Sumatra) *F. nivea* and *F. stipularis* are endemic. Other species occurring in this area are *G. crassifolius, S. ferrugineus, S. littoralis, S. macropterus, S. ridleyi*, and *X. fruticulosum*.

17. The ***west coast of Sumatra*** (Sumatra coast) and its islands are largely the same as the Woyla terranes of Metcalfe (Chapter 3). These terranes rifted in the Late Jurassic from the Gondwana margin towards the Eurasian one.
Only *S. ferrugineus* and *X. ferrugineum* are found here.

18. ***West Java*** (W Java) has long been part of the continent under which the Indian Plate was subducting. The remaining part of Java is Cenozoic volcanic rock on marine basement, that developed as result of the eastward extending volcanic arc. *Genianthus macrophyllus* is present as an endemic in this area. Other species occurring here are *G. ellipticus, G. crassifolius, S. ferrugineus, S. littoralis*, and *X. fruticulosum*.

19. ***Southwest Borneo*** (SW Borneo) is in fact the oldest part of Borneo (Borneo Basement). It rifted from the Eurasian margin in Jurassic times due to the opening of the Proto-South China Sea.
In this area *X. johannis-winkleri* occurs as an endemic. Other species here are *F. brachybotrys, F. splendidissima, S. auritus, S. ferrugineus, S. hirsutus, S. latistipulus, S. macropterus, S. maingayi, S. persicinus, X. capitatum, X. olivaceum*, and *X. pubistylosum*.

20. The ***Semitau*** (extended towards the region around Kuching) has a similar history as SW Borneo, and collided with the Borneo Basement in the Cretaceous.
Endemics in this area are *X. involucratum, X. sessile*, and *X. setosum*. Other species are *F. brachybotrys, F. splendidissima, G. maingayi, S. apoensis, S. auricomus, S. ferrugineus, S. macropterus, S. maingayi, S. persicinus, S. viridis, X. olivaceum*, and *X. pubistylosum*.

21 & 22. ***Sarawak and the central mountain range of Borneo*** (Sarawak/C Borneo) & ***Northwest Borneo*** (NW Borneo) are part of the Rajang-Crocker accretion that was uplifted in the Early Miocene, but before was part of a Gulf of the Proto-South China Sea. Since Palaeocene times the other parts of Borneo (also including W Sulawesi) were already land areas.

Fordia brachybotrys, F. splendidissima, S. albus, S. auricomus, S. ferrugineus, S. hirsutus, S. latibractea, S. macropterus, S. multiflorus, S. oblongifolius, S. persicinus, S. viridis, X. borneense, X. brookei, X. capitatum, X. glabrum, X. olivaceum, X. pubistylosum, and *X. semiorbiculare* are present in both areas. In addition *G. ellipticus* and *X. cylindricum* occur in area 21, and *G. crassifolius, S. maingayi,* and *S. sanguineus* in area 22. Endemic in area 21 are *X. alopecurum, X. glomeratum, X. grandifolium,* and *X. minus.* In area 22 there are no endemics.

23. ***Northeast Borneo*** (NE Borneo) is suggested by Hutchison (1989; Chapter 3) to be partly a continental fragment (Segama block), rifted off the Chinese margin in the Oligocene.

Fordia brachybotrys, F. splendidissima, F. rheophytica, S. albus, S. ferrugineus, S. hirsutus, S. latistipulus, S. macropterus, and *X. capitatum* are present. Endemic in this area is *X. longipediculatum.*

24. ***Central East Borneo*** (CE Borneo) consists of the continental fragment Mangkalihat, and the Kutei and Tarakan basins. This area originated after the opening of the Makassar Strait in the Early Tertiary. The Tarakan basin was a Miocene delta. Mangkalihat is regarded as a continental fragment rifted off Gondwana in the Late Jurassic together with W Sulawesi, and contains the oldest fossils of Borneo.

Fordia brachybotrys, F. splendidissima, G. ellipticus, S. ferrugineus, S. hirsutus, S. latistipulus, S. macropterus, S. maingayi, S. multiflorus, S. oblongifolius, S. persicinus, S. sanguineus, S. viridis, X. pubistylosum, and *X. semiorbiculare* are present. Endemic are *G. valvatus* and *X. foliaceum.*

25. The ***Meratus*** mountains have been uplifted rather late. Up to the Oligocene this part was continuous with the basins at both sides. It has been continuous with the South China magmatic belt (Chapter 3: Fig. 3.16), and consists of Cretaceous rocks. Together with the Semitau and SW Borneo, this is the oldest part of Borneo. The other parts of Borneo all accreted in later times, although some parts have a considerable age (see Mangkalihat in area 24).

Species occurring here are *F. brachybotrys, F. splendidissima, S. ferrugineus, S. littoralis, S. persicinus,* and *X. capitatum.*

26. ***Palawan*** is part of the displaced Chinese margin that rifted away after the opening of the South China Sea in the mid-Oligocene (32 Ma). Part of the Philippines (Mindoro) has the same origin. Rifting stopped 17 Ma.

Here *S. ferrugineus* and *X. ferrugineum* are present.

27 & 28. ***Northern Philippines*** (N Philippines) & ***southern Philippines*** (S Philippines) are both part of a geological complex and active region. The geology indicates different origins of parts of the Philippines. Most of the southern part is formed before the Miocene as an arc at the southern edge of the Philippine Sea Plate, whereas Luzon formed part of an arc at the north side of the Celebes Sea–West Philippine

Sea Basin in the same plate. Part of Mindoro (see also Palawan) is part of the continental sliver that rifted from the Chinese continental margin.

Of the studied genera no endemic species are present in this region. Species occurring in area 27 are *G. ellipticus*, *G. rectinervis*, *S, maingayi*, *S, sanguineus*, and *X. ferrugineum*; in area 28 *F. brachybotrys*, *G. rectinervis*, *S. apoensis*, *S. maingayi*, and *X. ferrugineum* are present.

29. Sulawesi is geologically very complex. However, the groups studied are not well represented in this area and, consequently, no subdivision of Sulawesi can be made using the present data. West Sulawesi has been part of the east margin of Borneo, but rifted away after opening of the Makassar Strait. Other parts have been rifted from the Australian margin, and collided with the west of the island 10 Ma.

Only the widespread species *S. ferrugineus* and *G. ellipticus* occur on Sulawesi.

RESULTS

The results of the analysis with PAUP under assumption 0 are: 19 equally most parsimonious area cladograms with a length of 366 steps and a consistency index of 0.45 (h.i. = 0.55, r.i. = 0.66, r.c. = 0.3). The 50% majority rule consensus tree and the strict consensus tree of the 19 most parsimonious area cladograms are presented here in Figure 4.4; differences are mainly caused by the changing position of the following areas, more or less within their clade: 1) India/W Burma, 2) Mergui/Tenasserim (and Andaman Is), 3) the clade of Sumatra coast and Palawan, and N Sumatra. The percentages of the trees that support a particular branch are indicated in the cladogram, but this does not mean that they support all the branches at the same time.

Groups that can be recognised in the cladogram are:

1) A group of areas on the *continent of SE Asia*: Kerala, India, W Burma/E India, and Yunnan/N Thailand with as sister Indochina. India shows up with an empty branch, because there are no endemics in the area, and the species occurring in India are also present in other related areas. In the strict consensus and the Adams consensus tree this clade is collapsed, except for Yunnan/N Thailand and Indochina. The reason for this is that W Burma/E India can move around in this clade (Fig. 4.4). Mergui/Tenasserim and the Andaman Islands are connected to this continental SE Asian area-group, but in different ways. This is probably caused by the low number of taxa in these area.

2) A small clade of two areas on the *Isthmus of Kra*: S Thailand and Penang/Kedah. This clade is added to the base of the continental SE Asia clade.

3) A group of areas in *Borneo*: Meratus, SW Borneo, Semitau, NE Borneo, NW Borneo, CE Borneo, Sarawak/C Borneo. The area NW Borneo is inhabited only by species occurring in areas sister to this area; because it does not have endemics, it shows empty branches.

4) A group on the *Malay Peninsula* extending into Java: W Malaya, Singapore, E Malaya, SE Sumatra, W Java, and connected to this group the N Philippines and S Philippines as sister areas.

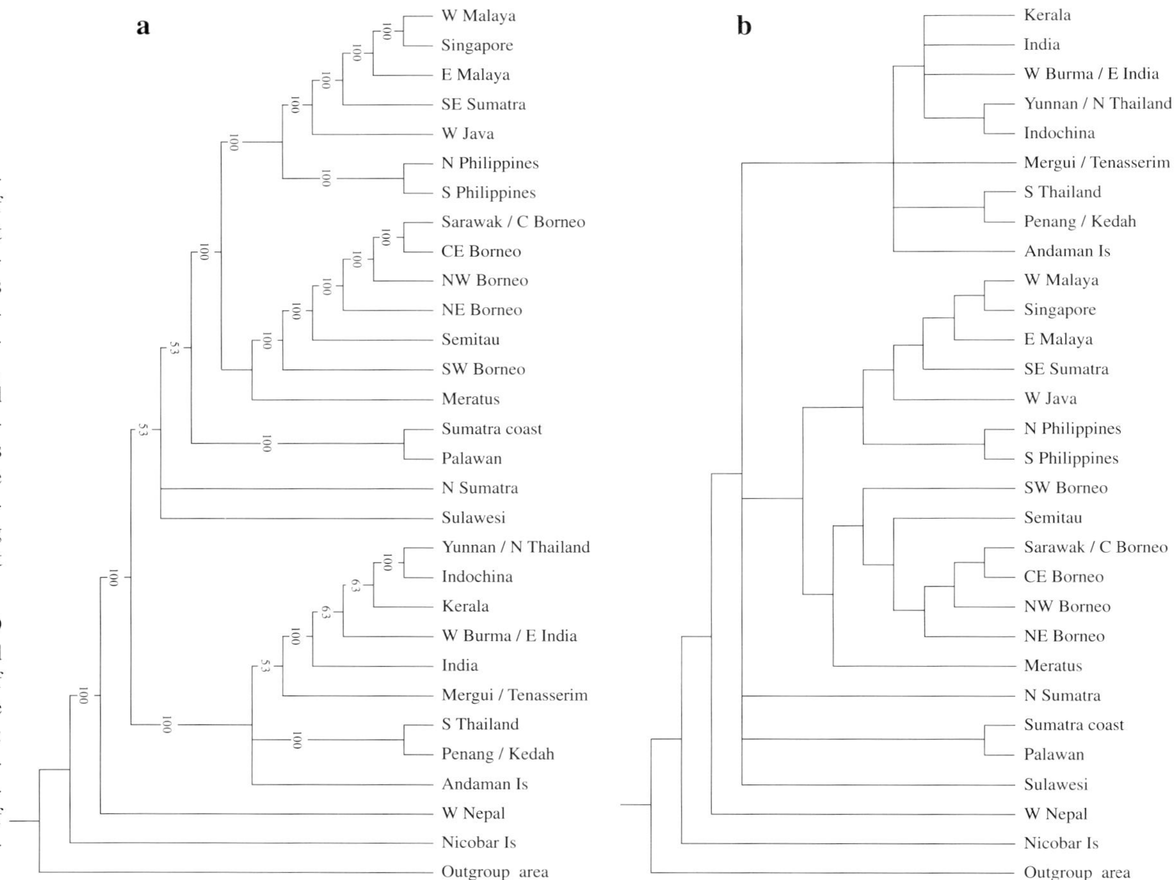

Figure 4.4. — **a**: 50% majority rule consensus tree of the nine-teen equally most parsimonious area cladograms obtained by analysis with PAUP of the area-taxon matrix of *Spatholobus, Butea, Meizotropis, Fordia, Genianthus,* and *Xanthophytum,* under assumption 0 (missing areas coded as 0). Values on the branches indicating the percentage of trees supporting a specific clade. — **b**: Strict consensus tree.

Differences between the 19 MPT's are mainly caused by the changing position of the following areas, more or less within their clade: 1) India/W Burma, 2) Mergui/Tenasserim (and Andaman Is), 3) the clade of Sumatra coast and Palawan, and N Sumatra.

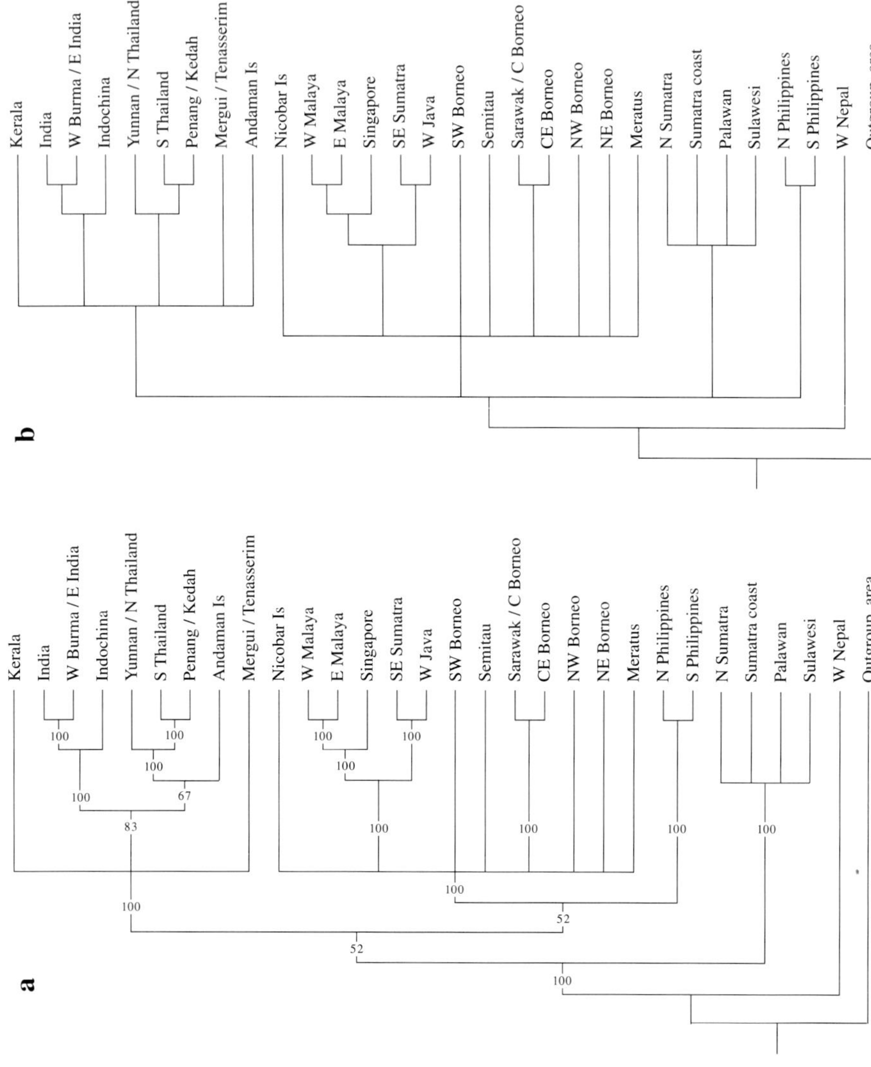

Figure 4.5. Consensus trees of all 324 equally most parsimonious area cladograms obtained by analysis of the area-taxon matrix of *Butea, Fordia, Genianthus, Meizotropis, Spatholobus,* and *Xanthophytum*, under assumption 1 (missing areas coded as ?). — **a**: majority rule consensus tree with indication of the percentage of trees supporting a specific clade. — **b**: strict consensus tree.

Some of the areas are placed low in the cladogram, probably due to lack of information: Nicobar Is, W Nepal, Sulawesi, Palawan, Sumatra coast, N Sumatra, and Andaman Is and Mergui / Tenasserim. The datamatrix shows, e.g., that on the Nicobar Islands only one species of *Genianthus* is present, and only one species of *Genianthus* and one of *Spatholobus* occur in Sulawesi. Palawan, Sumatra coast and N Sumatra are often connected to the Borneo and Malaya clades; the Andaman Is and Mergui / Tenasserim are often connected to the SE Asian continental clade.

Analysis under assumption 1 (missing areas coded as '?') resulted in 324 equally most parsimonious cladograms (length 316, c.i. = 0.52, h.i. = 0.48, r.i. = 0.69, r.c. = 0.36). The consensus of these 324 cladograms is shown in Figure 4.5. There are not as many clades resolved as in the results of the assumption 0 analysis. It is remarkable that the Borneo clade is nearly totally collapsed into polytomy.

The strict consensus tree (Fig. 4.5a) shows only resolved clades for:

1) Part of the *continental SE Asia group*: India as sister to W Burma / E India and Indochina; the rest of the group is a polytomy with the other areas mentioned above (the assumption 0 analysis).
2) The *Isthmus of Kra group* (S Thailand, Penang / Kedah) together with Yunnan / N Thailand.
3) The *Malay Peninsula group*: W Malaya, E Malaya, Singapore, W Java and SE Sumatra.
4) The complete *Borneo group* (but including the Nicobar Islands) has collapsed into a polytomy with the Malay Peninsula clade.

The aforementioned groups 1 & 2 and 3 & 4 are placed as a polytomy to the sister areas N Philippines and S Philippines, and the areas N Sumatra, Sumatra coast, Palawan and Sulawesi.

The 50% majority rule consensus tree is better resolved for the continental SE Asia group (Fig. 4.5b). The clades with 100% support are – naturally – the same as the clades for the strict consensus tree. The tree shows two major groupings:

1) A *continental SE Asia group* of India, W Burma / E India, Indochina, with as sister the group of S Thailand, Penang / Kedah, Yunnan / N Thailand, and the Andaman Islands. Kerala and Mergui / Tenasserim are both connected to the base of these clades. The place of the Andaman Islands, Kerala, and Mergui / Tenasserim is not certain.
2) The *Malay Peninsula group*: W Malaya, E Malaya, Singapore, W Java and SE Sumatra, in polytomy with the collapsed Borneo clade and the Nicobar Islands. The Philippines are added to the base of the group mentioned under 2. The unresolved clade of N Sumatra, Sumatra coast, Palawan, and Sulawesi, at the base of the cladogram, with W Nepal at its very base. This last group was also at the base in the PAUP area cladogram under assumption 0, and probably remains uninformative. W Nepal is basal to all other clades.

Figure 4.6. Generalised area cladogram obtained by analysis of the area-taxon matrix of *Butea, Fordia, Genianthus, Meizotropis, Spatholobus*, and *Xanthophytum* with CAFCA, under assumption 0 (missing areas coded as 0 in the datamatrix).

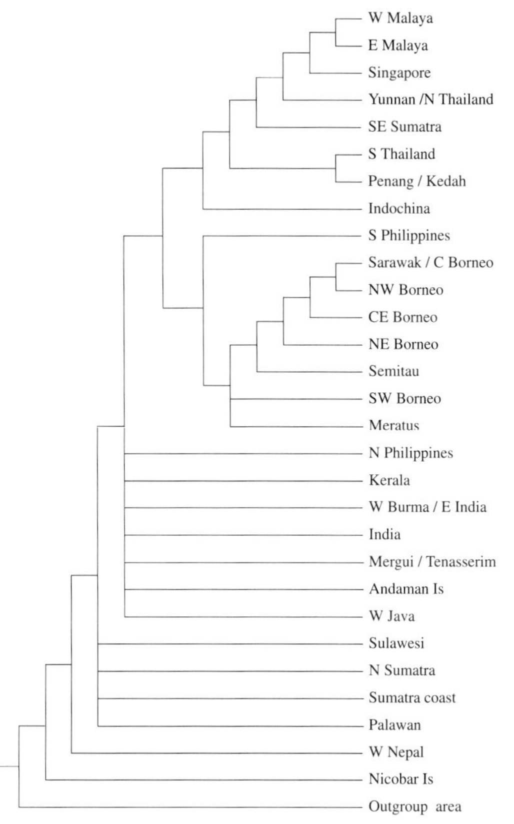

Analysis with CAFCA under assumption 0 leads to 37 cliques, out of which one is chosen by the minimal number of state changes (see Fig. 4.6; 450 steps, c.i. = 0.37, r.c. = 0.19; for comparison with the other results, length is 447 steps when calculated by PAUP). Here we see some of the already known groups again, with some slight changes:

1) The groups in *Borneo* (Meratus, SW Borneo, Semitau, NE Borneo, CE Borneo, Sarawak / C Borneo and NW Borneo), including the S Philippines at the base of the clade.

2) The *Malay Peninsula–Sumatra group* (W Malaya, E Malaya, Singapore, SE Sumatra, and in addition to the PAUP cladograms Yunnan / N Thailand at the base of Singapore) and the *Isthmus of Kra group* (S Thailand and Penang / Kedah), with Indochina at the base of these two groups, forming one clade.

3) The *continental areas in SE Asia*, i.e. Kerala, India, W Burma / E India, Mergui / Tenasserim, the Andaman Islands, together with W Java and the N Philippines are collapsed into one polytomy at the same level as the base of the two former groups together.

To the base of this, the configuration is the same as in the PAUP analysis under assumption 0: the unresolved clade of Palawan, Sumatra coast, N Sumatra and Sulawesi; at the base W Nepal; and then the Nicobar Islands at the very base.

Analysis with CAFCA under assumption 1 resulted in 1 most parsimonious tree out of 133 cliques found (length 417, c.i. = 0.396, r.i. = 0.68, r.c. = 0.48; Fig. 4.7). Now a slightly different cladogram is the result. Some groups are the same as in the cladogram obtained by the analysis under assumption 0:

Figure 4.7. Generalised area cladogram obtained by analysis of the area-taxon matrix of *Butea, Fordia, Genianthus, Meizotropis, Spatholobus*, and *Xanthophytum* with CAFCA, under assumption 1 (missing areas coded as '?' in the datamatrix.

1) The *Borneo group* (including also the S Philippines).
2) The *Isthmus of Kra–Malay Peninsula group* with Indochina at the base, but without Yunnan / N Thailand. Here, however, not as sister group to the Borneo one, but between the Borneo group and the rest.
3) The *continental SE Asia group* is more resolved than in the tree obtained with the CAFCA analysis under assumption 0 (W Burma / E India, Yunnan / N Thailand, India, W Nepal the Nicobar Islands, and Kerala). All other basally placed areas in the assumption 0 cladogram (Mergui / Tenasserim,

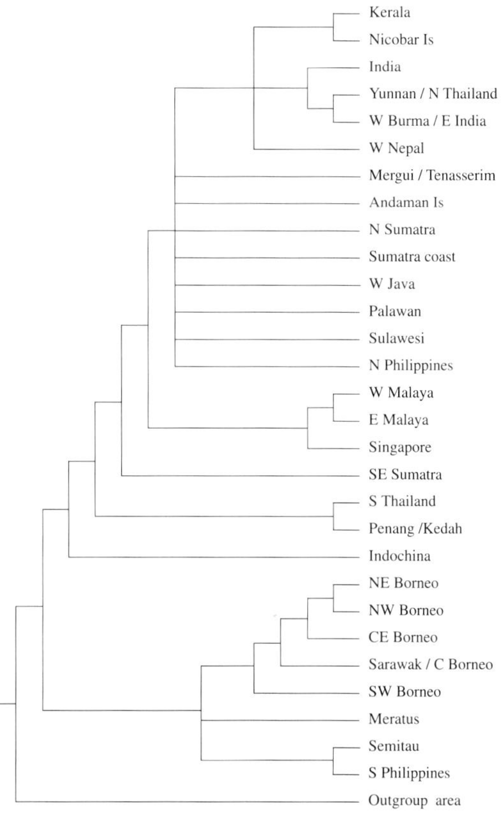

the Andaman Islands, N Sumatra, Sumatra coast, W Java, Palawan, N Philippines, and Sulawesi) are now in a polytomy at the base of the continental SE Asia group.

EVALUATION OF THE RESULTS

The results described above show that there are parts in all the obtained area cladograms that have similar topologies: a continental SE Asia group (unresolved under analysis with CAFCA), a Borneo group (in CAFCA including the S Philippines as well), an Isthmus of Kra group and a Malay Peninsula–SE Sumatra group (in CAFCA these last groups together with Indochina at the base). The area cladograms obtained by analysis with PAUP are the best resolved ones. As suggested in Sosef (1994) the clades in the cladogram found by PAUP were checked on the absence of taxa. Fortunately, the problem inherent in BPA analysis (Van Welzen, 1989, 1992), the definition of groups of areas on the absence of taxa, did not occur here. Sosef indicates that it is better to use BPA (PAUP) than CCA (CAFCA) in case of many extinction or dispersal events. Probably this is the reason for the better resolution of the PAUP cladogram compared to the one obtained by CAFCA under assumption 0. The better resolution of the cladogram, however, cannot be the only selection criterion. Comparison with geological information is another possibility to discriminate between the four possibilities presented above, although one has to be careful for circular reasoning in case of setting up a historical biogeographical hypothesis.

In contrast to the other area cladograms, the place of the unresolved areas in the CAFCA assumption 1 area cladogram is near the top of the tree instead of near the base. The continental SE Asia group plus the unresolved part of the tree are placed in such a way that the group of the Malay Peninsula–SE Sumatra–Indochina becomes a paraphyletic base of this part of the tree; the Borneo group forms the other part of the tree. It is possible that this result is mainly due to the fact that those areas, with a lot of missing data, all received a '?', which is translated in these areas as present (a '1') in connection to species occurring in the continental part of SE Asia.

The following areas are not very informative due to a lack of species occurrences, and probably for that reason appear low at the base of the cladogram: Sulawesi, the Nicobar Islands, W Nepal, N Sumatra, Sumatra coast, Palawan.

On the Nicobar Islands only one endemic species, *Genianthus nicobarensis*, is present. This single species, however, is outnumbered by the other species in the clade to define a node in the area cladogram, and the area is placed basal in the area cladogram. The place of the Nicobar Islands in the assumption 1 area cladograms must be artificial as well: in the PAUP area cladogram it is placed within the group consisting of the areas in Borneo and the Malay Peninsula, where one would expect it to be considering the distribution of its closest relatives. Under CAFCA, however, it is placed as sister to Kerala. This is probably another artefact of the analysis under assumption 1.

For West Nepal the situation is different: this area is probably placed at this basal position due to the place of the species occurring in this area; the species present in this area (*Butea monosperma*, *Meizotropis buteiformis*, *M. pellita*, and *Spatholobus parviflorus*) are all basal in the phylogenetic cladogram.

COMPARISON WITH THE GEOLOGICAL INFORMATION

It is not possible to construct a cladogram-like structure of the geological information (Chapter 3), because in the geological history of Southeast Asia the areas did not split up, but accumulated in SE Asia from the Gondwanaland margin. In the case of dividing areas (vicariance events), it is possible to construct a geological 'cladogram' in which the branches represent the geographical entities. The splitting up in this case, however, took place at the Gondwanaland margin, after which rifting occurred. The continental fragments rifted in different eras towards what is now SE Asia, and welded together into one large area (China/Indochina etc.) or some smaller parts (Sumatra, Borneo) within the SE Asian area. Below, these splitting and welding events are summarised; the different continental fragments and sutures are shown in Figure 3.5.

Before the Eocene no reliable fossils are known for the Leguminosae, and one cannot expect to see vicariance events before that time in this study. Only since Miocene times have fossils been found that can be attributed to specific genera within the Leguminosae. This is probably the reason why the areas of distribution on the continent and

Indochina (e.g., area 4, 5, and 6 in Fig. 4.3) are less similar to the geological entities than those in, e.g., Borneo, because they had already welded in the Triassic and formed one area at the time the Leguminosae developed.

Before the Eocene most of the SE Asian region was already present, but tectonic movement was still going on. The Tethys Ocean was subducting below the SE Asian margin that formed a volcanic arc from Lhasa, W Burma, Sumatra and part of Java to W Sulawesi, where the arc extended northwards. In the Early Eocene India started to collide with Asia. From the Oligocene to the mid-Miocene the South China Sea opened, and rifting from the Chinese continental margin towards the south took place. The Andaman Sea Basin and Thai Basin opened, and the Malay Peninsula rotated. Several changes took place in Borneo. In the Palaeocene large parts of Borneo were not yet uplifted, but an extensive landmass south and southeast of what is now Sarawak and Sabah was present, up to what is now W Sulawesi. Sarawak and Sabah were part of the Rajang Basin, facing the Pacific Ocean. The Crocker range represents an accretionary complex and deltas were present receiving sediment from the uplifted parts of Borneo as well as from the Indochinese part of the Sunda shelf. In Late Oligocene times, large parts of Borneo were inundated during a marine transgression phase. The Makassar Strait opened in the Late Eocene/Oligocene resulting in the separation of W Sulawesi. The rest of Sulawesi arrived at the end of the Oligocene. The Meratus mountains were uplifted rapidly in the Late Miocene-Pliocene. In the Miocene the Philippines were formed.

Until the mid-Miocene there were several periods of lower sea level. In the Late Oligocene, the sea level was as low as 250 m below the present one, thus the Sundaland plateau was for the largest part above sea level (see Fig. 3.4). The climate at that time was probably tropical monsoon, and before that, in the Oligocene, there was a wet tropical monsoon or a tropical rain forest climate. In the mid-Miocene the sea level changed again in three stages up to 220 m above sea level, after which there was a lowering again of up to 220 m below the present sea level in the Upper Miocene. At that time the Leguminosae became dominant among the phanerogams in India. According to Awasthi (1992, see also Chapter 3) this is after migration from Africa and SE Asia to India. In the Pliocene the sea level rose up to 140 m above the present level. After the Pliocene there were many glacial periods and interglacials which coincided with lower and higher sea levels, although by then the spectacular sea-level rises and falls were over (Hutchison, 1992).

If the groups studied were present in the region at the time these events took place and if they did react to these events, one may expect that this will be reflected in a general area cladogram.

The first major split in the general area cladogram that is obtained by analysis with PAUP assumption 0 is between the *continental SE Asia group* (inclusive the Isthmus of Kra group and the Andaman Islands) and the *Sundaland groups* (Malay Peninsula/ Sumatra and the Borneo areas). This may be due to one of the high sea levels, which occurred between the mid-Eocene and the Pliocene as indicated above (see also Table 3.1). It is clear from the reconstructions of Rangin et al. (1990b), that during high sea level there was a separation between the Malay Peninsula and Indochina. It is impossible to indicate more exactly in what period this splitting off took place. If the place of the Andaman Islands is correct, i.e. below the split between continental SE Asia and the Sundaland groups, it may be possible to date the split of the Andaman Islands to the time of the opening of the Andaman Sea (13 Ma). This is very speculative, however, because on the Andaman Islands only the widespread, but continental, species *Spatholobus acuminatus* is found. It would be better (for future research) to add a genus to the analyses with one or more endemics on the Andaman Islands.

All area cladograms, except the one obtained by PAUP assumption 1, are resolved for the areas in *Borneo*. In the area cladogram obtained by PAUP assumption 0 (Fig. 4.4a), the first area to split off is the Meratus. It is possible that this reflects the uplift of the Meratus mountains in the Late Miocene-Pliocene. The second area is SW Borneo, which is the oldest part of Borneo. The third area is the Semitau, which is in fact the extended part of the Semitau ridge, which was accreted to SW Borneo in the Cretaceous. This part of Borneo is probably uplifted in the same period as the Meratus (Hutchison, 1992). The next area, NE Borneo, was, according to Hutchison (1992), also part of the extensive landmass, and uplifted in the same period as the Meratus. This implies that the first four areas in the Borneo group were not yet uplifted up to the Late Miocene-Pliocene, and were probably submerged during some of the sea level rises. The first four splits (in both assumption 0 general area cladograms) may indicate these uplift events and the related changes in climatological/ecological factors. The other parts of Borneo in these times consisted of the large Rajang Basin and the Crocker Range (Sarawak/C Borneo and NW Borneo). The last split in the Borneo group (PAUP assumption 0) is formed by the central parts of Borneo: Sarawak/C Borneo and CE Borneo. The mountains of the Crocker Range form a natural delimitation between CE Borneo and the west central part of Borneo (Sarawak/C Borneo).

The *Malay Peninsula* group is another group that can be found in most area cladograms. W Malaya and E Malaya form in most cases the top of this clade together with Singapore. SE Sumatra is at the base. In the PAUP area cladograms W Java is basal in the same clade, and the N and S Philippines are connected to the base as well. In the area cladograms obtained by CAFCA, W Java and the N Philippines form part of the polytomy with the unresolved SE Asian areas; the S Philippines are placed basal to the Borneo group. There is no difference between the area cladograms in the sister relationship between the E Malaya part of the Indochina block and the W Malaya part of the Sibumasu block; these have a sister relation to each other in the area cladogram together with Singapore. SE Sumatra belongs to the Sibumasu block. However, the connections between these blocks date back to the Triassic, long before the Leguminosae existed there. Although it is possible to recognise different distribution areas, more or less according to the delimitations of Sibumasu and E Malaya, it is not possible to distinguish in the area cladogram between these areas. The delimitation of the distribution areas may of course be due to ecological or climatological factors as well: the area around Singapore and SE Sumatra are both lowland areas. The E and W Malaya parts of the Malay Peninsula consist largely of the large, mountainous central area. If there is a relation between Malaya and Sumatra, one would expect, on account of distance, that there would be as many species occurring in W Malaya and the mountains of NW Sumatra as there are species in the lowlands around Singapore and SE Sumatra. From the distribution of the species in this study the former is not evident. No species occurs exclusively in both W Malaya and N Sumatra. The relation between the Malay Peninsula (West, East and Singapore), SE Sumatra and W Java may be understood in geological terms also as part of the border of Sundaland below which the Indian Plate was subducting. Only after the rotation of the Malay Peninsula, the northward movement of Burma and the opening of the Andaman Sea (13 Ma) were these parts (slightly) rearranged. The reconstructions in colour of Rangin et al. (1990b) indicate that during certain periods of high sea level, between the Middle Eocene and the Pliocene, there were emergent areas in Borneo, the Malay Peninsula, large parts of Sumatra, and West Java (Fig. 3.4). Other parts are shown as submerged, although based on continental crust, e. g., the areas S Thailand and Penang / Kedah. The S Thailand and Penang / Kedah areas are placed very close to each other in all area cladograms. In the PAUP assumption 0 analysis this clade is in polytomy to the continental SE Asian areas and the Andaman Islands. In the CAFCA analyses the same clade is found at the base of the Malay / Sumatra group (with Indochina). They form the region around the *Isthmus of Kra*, which was easily inundated during periods of high sea level (Fig. 3.4). Floristically this is the place where the distribution of the species from everwet and seasonal areas overlap. The Malay Peninsula group as indicated by the PAUP assumption 0 area cladogram, which also included WJava at the base of the group, fits this reconstruction best. It is probable that the area around the Isthmus of Kra was later invaded, an assumption supported by the fact that this small group of two areas does not contain any endemic species. It is possible that rain forest species do not have as much potential for dispersal than monsoon forest ones, and it is then probable that the invading species came mainly from the north and less from the south (with its rain forest climate). In this view, the place of the Isthmus of Kra clade basally to the continental SE Asia areas is the best explanation, as is found in the area cladogram obtained by PAUP assumption 0.

The fact that the N and S Philippines are placed closely to each other favours the PAUP assumption 0 area cladogram. The place of the Philippines as close to Borneo as it is to West Java (and the rest of the Malay / Sumatra clade) perhaps indicates that Borneo had its own history, separate from the Philippines.

On the basis of the best resolved tree and the geological comparisons I give preference to the PAUP assumption 0 area cladogram, especially because of the relations within the Malay / Sumatra group (no intruding continental areas), the resolved part of the continental SE Asia group, the place of the Isthmus of Kra clade.

RECONSTRUCTING THE HISTORY OF SPATHOLOBUS

As shown above, the differences between the PAUP and CAFCA analyses are mainly in the unresolved part of the area cladogram, where either the continental SE Asian areas (CAFCA) or the Borneo areas (PAUP assumption 1) are unresolved. In order to reconstruct the possible history of speciation within *Spatholobus*, the phylogeny of the genus *Spatholobus* and allies (Fig. 4.1) is optimised on the generalised area clado-

Figure 4.8. Generalised area cladogram obtained by analysis with PAUP of the area-taxon matrix of *Butea, Fordia, Genianthus, Meizotropis, Spatholobus*, and *Xanthophytum*, under assumption 0 (= missing areas coded as 0 in the datamatrix). Species of *Butea, Meizotropus*, and *Spatholobus* are superimposed on the area cladogram. ● = present; x = absent.

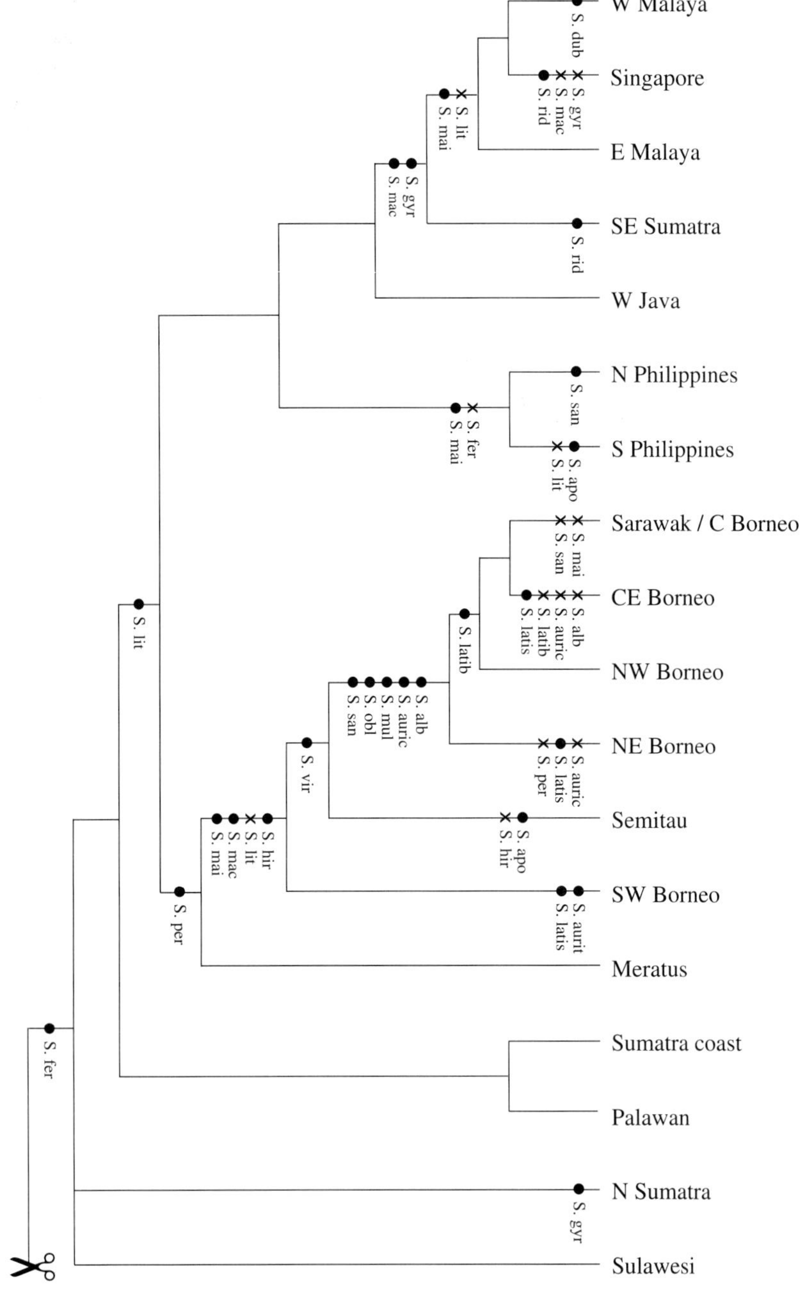

(continued from page 118)

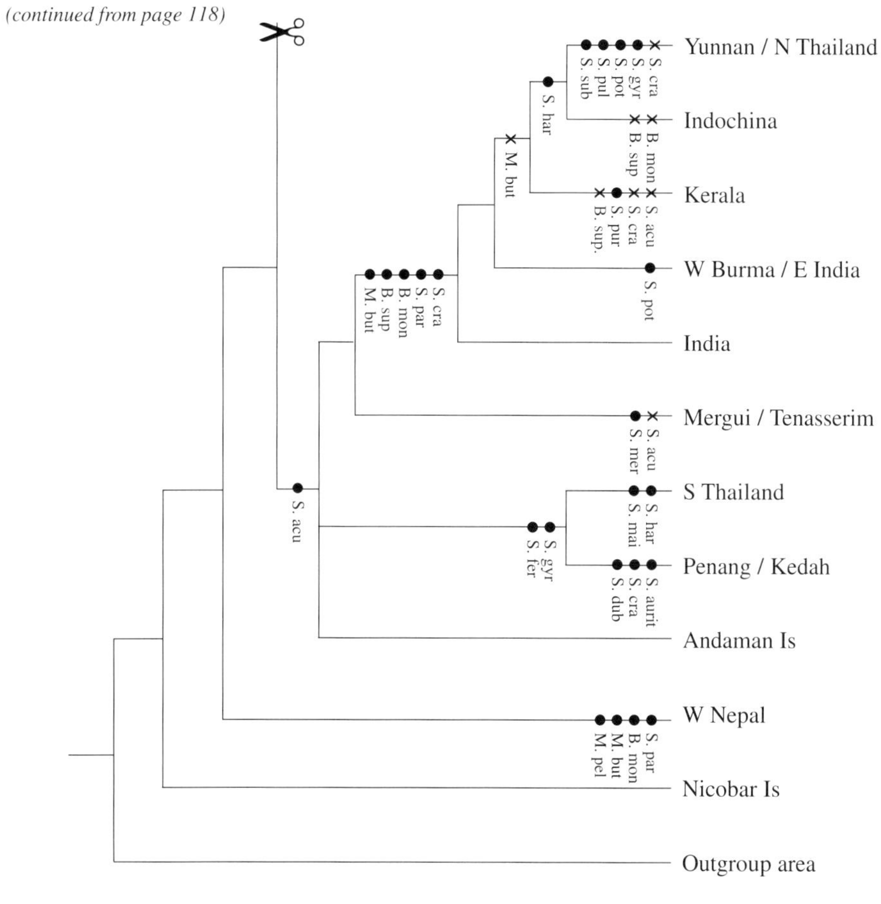

gram that is the result of the biogeographical analysis with PAUP under assumption 0 (Fig. 4.8). The speciation events indicated by the cladogram of *Spatholobus* will be used to relate the general area cladogram to the phylogenetic history of the genus as reconstructed in the *Spatholobus* cladogram. It is assumed that the phylogenetic cladogram and the general area cladogram correctly reflect the history of speciation of the genus and the historical relation of the areas respectively.

Below, the implications of the optimisation on the area cladograms is described. The most representative parts of the cladogram of *Spatholobus* and corresponding parts of the area cladogram will illustrate this. When in the text an ancestor (A) is mentioned, a number (in parentheses) is given indicating its place in the cladogram (Fig. 4.1a). The basal – for the greater part polytomous – areas, however, are mostly excluded, i.e., N Sumatra, the west coast of Sumatra, Palawan, W Nepal, the Nicobar Islands and Palawan.

In all area cladograms *Spatholobus* (and also *Butea* and *Meizotropis*) is absent from the Nicobar Islands; in the PAUP assumption 0 area cladogram this is considered to be a primitive absence, in the other area cladograms the option of extinction is also open.

The second area that splits off in the area cladogram is W Nepal. Of the genus *Spatholobus*, only *S. parviflorus* occurs here. If this presence is due to dispersal from, e.g., India, the rest of *Spatholobus* can be assumed to be originally present (i.e., the ancestral area of the genus) in all four area groups (continental SE Asia, Malay Peninsula/ SE Sumatra, Isthmus of Kra and Borneo).

The areas at the base of the Sundaland clade (N Sumatra, the SW coast of Sumatra, Palawan and Sulawesi) are inhabited by only one species of *Spatholobus*, the widespread *S. ferrugineus*. The closely allied *S. gyrocarpus* occurs also in N Sumatra, but not in the other three areas mentioned.

The ancestor of *Butea* and *Meizotropis* (A45) is present on the continent of SE Asia, but not in Indochina or Mergui/Tenasserim (Fig. 4.9). It is also found in W Nepal with its area placed low in the cladogram. It is possible that W Nepal split off early, and that the other species never reached this region because the area became isolated by the extreme climatic circumstances. *Meizotropis* (A44) has a more restricted distribution than *Butea*, and probably originated in India and W Burma/E India (in fact the actual area will be even more restricted, but these are the areas delimited for this analysis).

The ancestor of *Spatholobus* (A47) is optimised in the area cladograms as occurring in the whole area, except the Nicobar Islands (Fig. 4.9). The most basal splits, however, are the ones of *S. parviflorus*, and those of the genera *Butea* and *Meizotropis*, all occurring only on the SE Asian continent. Because all later clades have ancestors optimised as occurring one way or another in both the SE Asian areas and the areas southwards and including the Isthmus of Kra, the ancestors of these first splits (*S. parviflorus*, *Butea* and *Meizotropis*) are optimised as occurring in the whole area as well. In this option, the parts of Borneo and the Malay Peninsula became isolated later, providing opportunities for geographical differentiation in these isolated areas. It is, however, not necessary that the first ancestor occurred in the whole area. It is easily possible that the limits of the area were those of the present distribution of *S. parviflorus*,

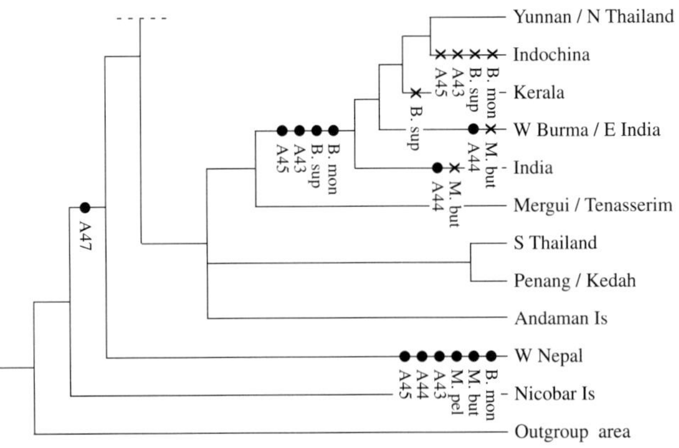

Figure 4.9. Part of the area cladogram (PAUP, assumption 0) showing *Butea, Meizotropis* and the ancestor node of *Spatholobus* (A47). ● = present; x = absent.

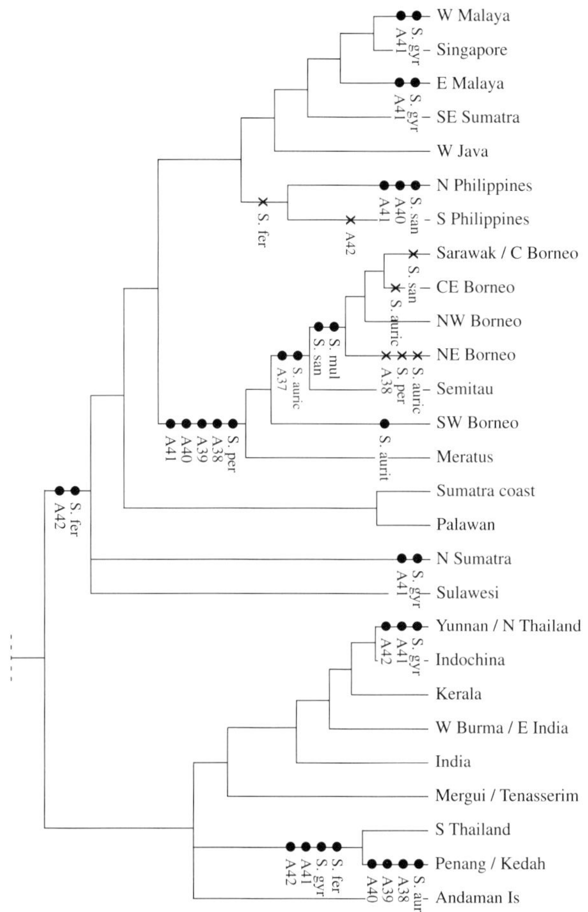

Figure 4.10. Area cladogram showing the *Spatholobus ferrugineus*-clade. ● = present; x = absent.

and that at a certain time the genus *Spatholobus* dispersed over the whole area, when the circumstances were favourable (e.g., low sea level during Late Oligocene). If one would wish to date the ancestor of the group much earlier, the group could have had a distribution including SW Borneo, because in Early Eocene times there was an emerged land mass connecting the SE Asian continent to what is now SW Borneo. In Middle Eocene times, the high sea level already separated Borneo from Sumatra and the Malay Peninsula.

The ancestor of the *S. ferrugineus*-clade (A42), the second split in the cladogram, can be found in all of the Borneo areas, the Malay Peninsula/SE Sumatra areas (including W Java), the Isthmus of Kra, and Yunnan/N Thailand (Fig. 4.10). In the CAFCA results Yunnan/N Thailand is incorporated in the Malay Peninsula/SE Sumatra group of areas, whereas the analysis with PAUP resulted in combining Yunnan/N Thailand with the continental SE Asia areas, and in the Malay Peninsula/SE Sumatra group including W Java (instead of sending it into polytomy). If the total of extant areas is indeed reflecting the ancestor area, it will be most parsimonious to assume that both W Java and Yunnan/Thailand belong to the Malay Peninsula/SE Sumatra group of areas. If, however, part of the area has been reached by one or more (ancestor) species in later times by dispersal, it is very well possible that, e.g., Yunnan/Thailand is part of the continental SE Asia group of areas instead of the Malay Peninsula/SE Sumatra area group. However, the fact that the ancestor of the *S. ferrugineus*-clade is present in the area Yunnan/Thailand, is due to the presence of only one specimen of *S. gyrocarpus*, a species that occurs otherwise only south of the Isthmus of Kra. This could be explained by an occasional dispersal of this species to the north in more recent times. In the latter case the result of PAUP reflects the history better than the result of CAFCA. On the other hand, if dispersal from the ancestral area into W Java took place, it is possible that the CAFCA area cladogram gives a better solution. The ancestor of the *S. ferrugineus*-group is shown otherwise absent from the continental SE Asia areas and present in the whole of the areas from the Isthmus of Kra southwards. This indicates that rather early in the history of this group there has been a vicariance event between the continent of SE Asia and the other parts of its distribution area. It is possible that after a period of isolation, i.e. high sea levels (Late Eocene/Early Oligocene, mid-Miocene, Pliocene), speciation took place. In the palaeogeodynamic maps of Rangin et al. (1990b) it is obvious that during all periods of higher sea level the Isthmus of Kra region was inundated, thus leaving a smaller or larger gap between the Malesian region and the continent of SE Asia (Fig. 3.4). The *S. ferrugineus*-group, however, either never reached the S Philippines, or it went extinct later. In the CAFCA area cladogram the N Philippines are not included in the Borneo-Malay Peninsula groups, but are placed lower in the area cladogram, and in polytomy. It is possible that the occurrence of *S. sanguineus* in the N Philippines (the only species in the *S. ferrugineus*-group occurring here) is due to dispersal in later times. In this case the *S. ferrugineus*-group is primitively absent from the Philippines, and may even have been present in the area before the Philippines arrived. This would imply at least a Middle to Late Miocene age for the ancestor of the *S. ferrugineus*-group. It is also possible that the ancestor never reached the Philippines, although these islands were already present in the region. However, it is hard to believe that such a widespread species as *S. ferrugineus* did not disperse over, e.g., the whole island of Java, or from Palawan into the Philippines. It is more satisfactory in this case to invoke climatological or ecological factors to explain the absence of *S. ferrugineus* than to explain it by 'primitive' absence (because it could not reach the Philippines). The Philippines, and also the more eastern and central parts of Java have a relatively dry climate, compared to the everwet rain forest climate (e.g. of Borneo). It is possible that the ancestor of *S. ferrugineus* became extinct in the Philippines, and that the later developed *S. sanguineus* is better adapted to drier circumstances, although it is occurring in the wet parts of Borneo as well. *Spatholobus gyrocarpus* developed, as second species in this clade, on the Malay Peninsula and N Sumatra. The barrier between Borneo on the one side and the Malay Peninsula and Sumatra on the other was probably too large to bridge. *Spatholobus gyrocarpus* occurs in the higher parts of the Malay Peninsula, an area that was continuous with the higher parts of N Sumatra during periods of low sea level. *Spatholobus gyrocarpus* is absent from the lower parts of the Malay Peninsula (towards Johore and Singapore) and from SE Sumatra. *Spatholobus sanguineus,* on the other hand, probably developed on the northern and central part of Borneo. These are parts that later accreted to the older parts of SW Borneo, Meratus and Semitau. According to Hutchison (1992) these older parts formed a single landmass continuous with W Sulawesi and the Java Sea in the Tertiary. In Early Miocene times the accreted sediments of Sarawak and the Crocker Formation were compressed and uplifted. The younger parts were uplifted in the Late Miocene-Pliocene after having been filled with sediments. This is also the case for *S. sanguineus*, which occurs in areas that contain the higher parts of – in this case – Borneo and the northern Philippines. It is not evident from the specimen labels, however, that the species themselves occur only in higher altitudes; on the contrary, the only indication of altitude for *S. gyrocarpus* is up to 160 m above sea level; for *S. sanguineus* the altitude is indicated from low altitudes up to 1000 m. The fact that *S. sanguineus* is found on the northern Philippines indicates either a vicariance between Borneo and the Philippines (a more recent dispersal event), or a primitive presence of the clade in the Philippines. In the latter case *S. sanguineus* is a remnant of a once larger distribution of the *S. ferrugineus*-clade. Consequently, the ancestor of the *S. ferrugineus*-clade was already present in the area in Late (or even Middle) Miocene times, and *S. sanguineus* developed in the newly accreted and uplifted areas in the northwest of Borneo and the Philippines, which came within reach especially after a lowering of the sea level in the Late Miocene. The Philippine group of *S. sanguineus* has already been separate from the Borneo part of the species for a considerable period of time, indicated by slight morphological differences, e.g., the flowers of the Philippine specimens are larger than those of Borneo, which have very small flowers. *Spatholobus auricomus* and *S. multiflorus*, two other species with small flowers within the *S. ferrugineus*-clade, also developed on the northern half of Borneo (Sarawak/C Borneo, CE Borneo, NW Borneo, and NE Borneo). In addition to the northeastern parts of Borneo, *S. auricomus* also occurs in the Semitau area, which is close to Sarawak. This species does not occur in the eastern part of the northern half of Borneo (CE Borneo and NE Borneo), and only in the southernmost part of the NW Borneo. *Spatholobus multiflorus* occurs in the complementary parts. It is probable that the mountainous area of C Borneo is the natural delimitation of *S. auricomus* and *S. multiflorus*. Both species do not occur in the parts with the highest altitudes. The ancestor probably occurred in the whole area, and the two species developed on either side of the central mountain part. It is, however, difficult to give an indication of timing for this last event. Sarawak and the Crocker Formation were uplifted in the Early Miocene, and these mountains were probably already in existence before the ancestor of the species came into the region. The parts south of the mountains were inundated during the Late Oligocene. The high sea level that isolated the Meratus from the rest of Borneo during the Pliocene may also have played a role in the differentiation of these two species. The other split at the end of the *S. ferrugineus*-clade is that of *S. persicinus* and *S. auritus*. *Spatholobus persicinus* occurs in the whole of Borneo, but *S. auritus* is found only in SW Borneo and in Penang/Kedah. Apart from *S. ferrugineus* itself, *S. auritus* is the only species that occurs both in Borneo and on the Asian mainland. The species is not

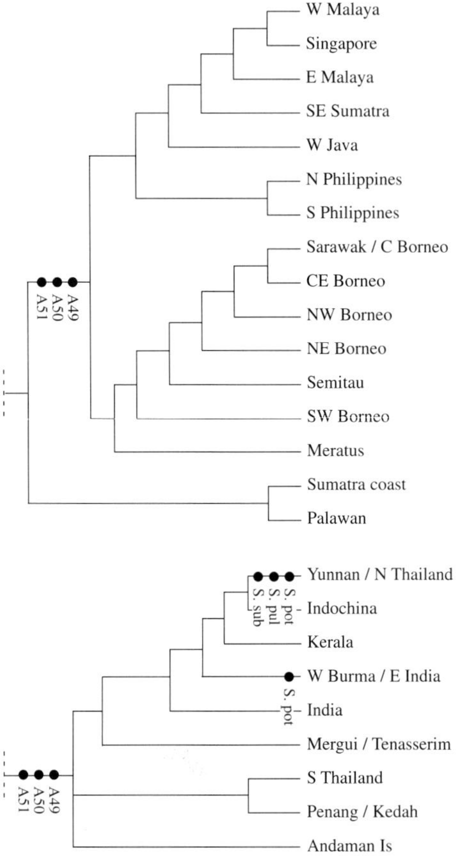

Figure 4.11. Area cladogram showing the species *Spatholobus pottingeri*, *S. pulcher* and *S. suberectus*.

very common, and except for one (sterile) specimen in Borneo there are indications that it is in fact endemic in Penang. The occurrence of *S. auritus* in Borneo suggests that it is either a relict of a more widespread distribution, or a dispersal event, or an identification error (the specimen is sterile).

The next split off in the *Spatholobus* cladogram is that of *S. pottingeri*, occurring in Yunnan / Thailand and W Burma / E India (Fig. 4.11). This species probably came into existence at the periphery of the ancestral area of *Spatholobus*. The Yunnan / N Thailand area remains present in the ancestral areas of the next two splits (A50, A51). The first split is that of *S. pulcher*, an endemic occurring in the higher parts of the area Yunnan / N Thailand. After a period of being more or less isolated, the area probably became occupied by the ancestor of *S. pulcher*. In the Pliocene, tropical montane forest was present in the area (Guo, 1993). *Spatholobus pulcher* grows between altitudes of 1325 and 1525 m. The second species, *S. suberectus*, is found also at high altitudes (1250–2500 m) in evergreen forest. It is possible that by a climatic change this area became isolated, and that the ancestor of *S. suberectus*, occurring at the periphery of the whole area of *Spatholobus* on the continent, became adapted to living at high altitudes.

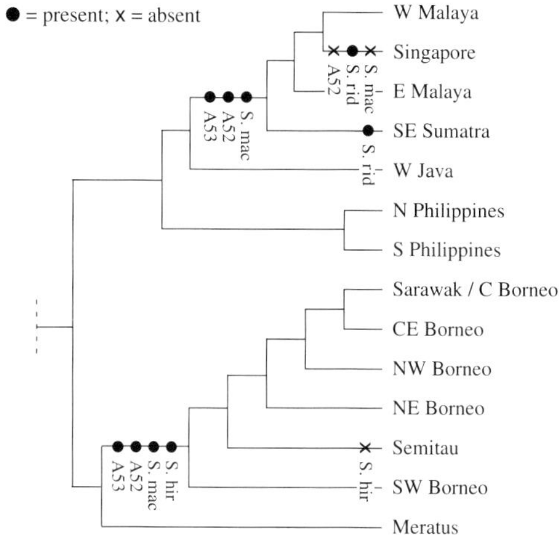

Figure 4.12. Area cladogram showing the *Spatholobus macropterus*-clade.

The next clade, that of *S. macropterus*, is present in the whole Sundaland area, except W Java, the Philippines, and the Meratus area on Borneo (Fig. 4.12). It is likely to assume a primitive absence in these areas. The first split off from its ancestor (A53) is *S. ridleyi*, occurring only in the south of the Malay Peninsula (Singapore area) and SE Sumatra. *Spatholobus hirsutus* is present on the whole island of Borneo except for the Meratus. As far as information on the labels allows, *S. hirsutus* is present at low altitudes. *Spatholobus macropterus*, which occurs nearly in the whole area of the clade, is found up to 1700 m.

The next species to split off from the ancestor is *S. oblongifolius*. This species occurs in the northern, more recent accreted parts of Borneo.

The ancestor of the *S. maingayi*-clade (A36), the next split in the cladogram, is present in the Sundaland areas (Fig. 4.13). The ancestor of the whole genus (A56) is assumed to be still present in the whole area. Ancestor 36 occurs in the whole region south of the Isthmus of Kra, including S Thailand. The first species that developed is *S. littoralis*, which occurs in a rather restricted area: the Meratus in Borneo, SE Sumatra, W Java, and the northern Philippines. It is best to assume that the species was restricted to these parts and thus primitively absent from the other parts. The next species splitting off

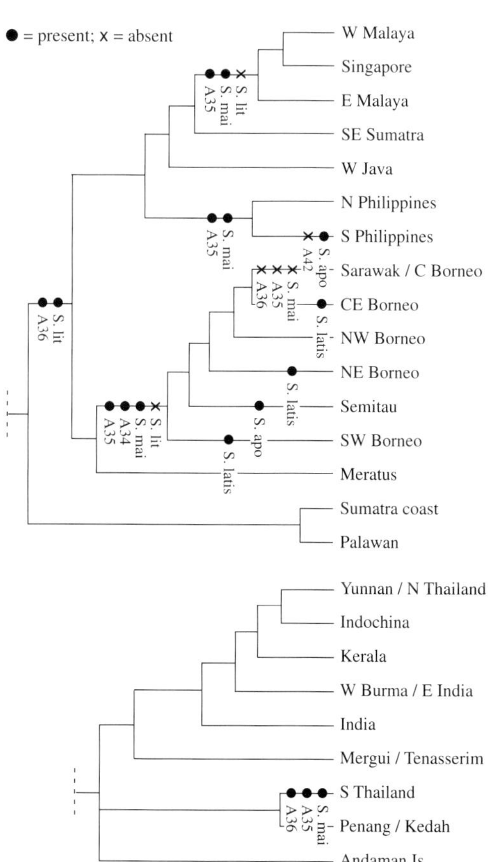

Figure 4.13. Area cladogram showing the *Spatholobus maingayi*-clade.

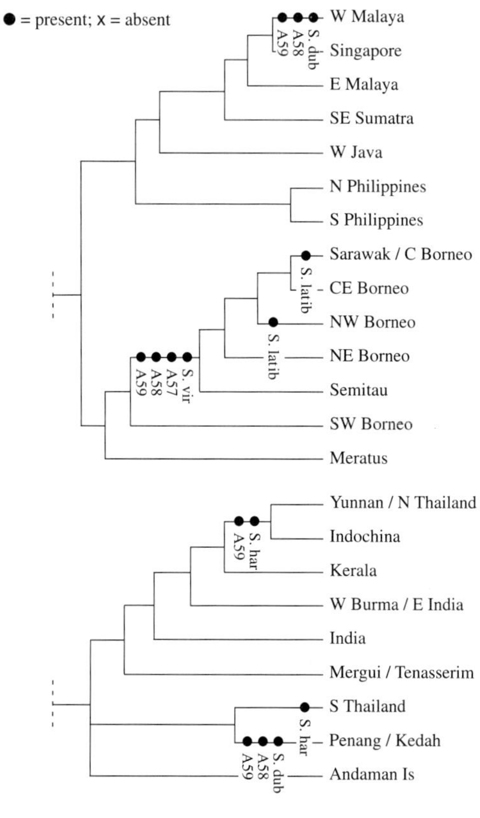

Figure 4.14. Area cladogram showing the *Spatholobus harmandii*-clade.

in this clade is *S. maingayi*, which is more widespread than *S. littoralis*, and except for W Java reflects the area of ancestor 36. *Spatholobus apoensis* and *S. latistipulus* occur in part of the area of *S. maingayi*: *S. latistipulus* only in Borneo (SW Borneo, NE Borneo, CE Borneo), and *S. apoensis* in the southern Philippines and the Semitau. This last species has a disjunct distribution, which is perhaps due to extinction.

The ancestor to the rest of *Spatholobus* (A60) does not occur in the oldest parts of Borneo and the Malay Peninsula, except for W Malaya. The ancestor of the *S. harmandii*-clade (A59) is present in both Yunnan/N Thailand and Indochina, the Isthmus of Kra region, W Malaya and the northern part of Borneo (Fig. 4.14). The first species in this clade, *S. harmandii*, is restricted to the area north of S Thailand. It is most likely that a vicariance event took place which isolated the ancestor of *S. harmandii* from the rest. If this event was a sea level rise, the Isthmus of Kra will have been inundated, and the occurrence of *S. harmandii* in S Thailand can then be explained by dispersal from the north. Obviously this species is adapted to a drier climate, as indicated by the thick leaflets, and was (and is) prevented from dispersing further south by the wet climate. The species south of the Isthmus of Kra survived in rather restricted

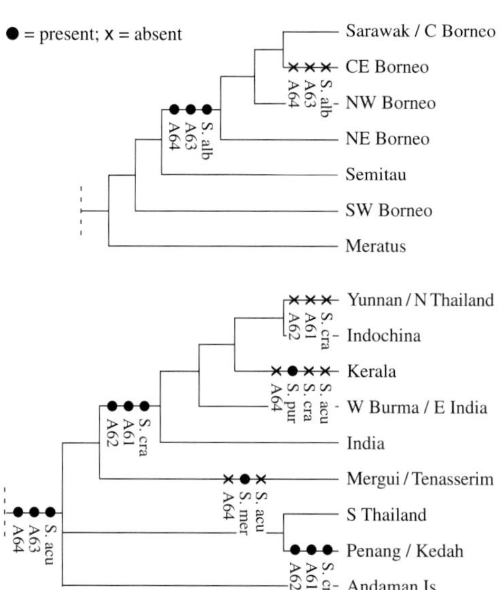

Figure 4.15. Area cladogram showing the clades of *Spatholobus acuminatus* and *S. classifolius*.

areas: W Malaya/Penang/Kedah and the northern parts of Borneo (Fig. 4.22). In West Malaya, *S. dubius* is present in only the part close to Penang/Kedah, and is nearly endemic. The presence of this species is in fact a relict of the larger ancestral area from Yunnan in the north over the south of Thailand into the Malay Peninsula and further to Borneo. In the Malay Peninsula no species of the *S. harmandii*-clade are found, and it is most parsimonious to assume that in these areas the ancestor went extinct, and survived only in Borneo and the Penang and surrounding areas. On Borneo two species occur: one up to 350 m altitude, *S. viridis*, and the other, *S. latibractea*, in a more restricted area at higher altitudes (1300 m).

The ancestor of the remaining part of *Spatholobus* (A63) is optimised in the area cladogram as occurring in the whole SE Asian region, and in the north and western parts of Borneo. Ancestor 64 occurs in a less restricted region than ancestor 62 (Fig. 4.15). Ancestor 64 gave rise to two species occurring in completely different areas: *S. albus* in the northern part of Borneo except for CE Borneo, and *S. acuminatus* in the area north of and including the Isthmus of Kra. This last species is (primitively) absent from Kerala and the Mergui/Tenasserim. Probably this split is as much the result of the differences in climate as that of isolation during high sea level. The area in which ancestor 62 occurred is, except for Yunnan/N Thailand and S Thailand, the whole of the continental SE Asia area. The first speciation, resulting in *S. merguensis*, took place in the Mergui archipelago that was separated first; then Kerala was split off (became isolated) resulting in *S. purpureus*. This last event was perhaps due to the remaining everwet climate, whereas in large areas of continental SE Asia the climate became drier. The remaining areas form the distribution area of *S. crassifolius* that is widespread in the rest of continental SE Asia.

During the whole development of the genus *Spatholobus* there must have been several events during which the Malay Peninsula/Sumatra or the Borneo region were isolated from the mainland of SE Asia. Events that could be held responsible for such an isolation can be found in the sea level changes followed by a change in humidity and probably temperature. Timing is very speculative, however, because there have been a considerable amount of sea level changes during all periods.

SUMMARY OF THE RELATIONS OF SPATHOLOBUS THROUGH SPACE AND TIME

It is rather speculative to superimpose older (= older than Pleistocene) geological events on the cladogram of *Spatholobus*. Although some events can be reconstructed, it is not possible to designate the time these events happened, especially because most events, i.e. changes in sea levels, have occurred several times during geological history. The best possible way to reconstruct fragments of the history of *Spatholobus*, with the available area cladograms and the present knowledge of geology, is to assume that a taxon has to come from somewhere in the first place (distribution of the ancestor) and then find explanations on the deviations from this area. Later events, e.g. Pleistocene ice ages, can easily have disrupted part of the distribution area of some of the species and their ancestors, resulting in disjunct distribution patterns, which in their turn may have resulted in speciation events. However, as not all species will react on separation with speciation (non-adaptive radiation), some of these disjunct patterns will remain.

The generalised area cladograms indicate a vicariance event between the areas on the continent of SE Asia and the areas south of the Isthmus of Kra. It is probable that this indicates one of the sea-level rises in the Late Eocene, the mid-Miocene, or the Pliocene. During all sea-level rises the areas between Borneo, Sumatra and the Malay Peninsula were inundated. The Isthmus of Kra region was also frequently inundated. In this area no endemics of the groups under study occur, only species that dispersed from either north or south towards this region. The area forms an overlap between continental SE Asia and the West Malesian areas.

In the area south of the Isthmus of Kra two separate groups of areas can be recognised: Borneo and the Malay Peninsula / SE Sumatra. There is a relation between SE Sumatra and W Java on one hand, and SE Sumatra and E Malaya on the other. West Java has long been part of the continental margin under which the Indian Plate subducted; SE Sumatra and E Malaya were part of the continental fragment that included Indochina as well. West Malaya, the region round Singapore, and E Malaya form one part. This is, however, not reflecting the older geological pattern of a separate Sibumasu and Indochina / E Malaya that welded in the Triassic. In Borneo the geological entities are recognisable in part in the distribution areas of the species; the oldest parts are the ones split off first, and the younger parts in the north are inhabited usually by the younger species. The uplift in the Late Miocene-Pliocene of parts of Borneo, e. g., Semitau and the Meratus, is probably reflected in the first splits.

The ancestor of the genus *Spatholobus* probably originated on the SE Asian continent. The genera *Butea* and *Meizotropis* developed in that part of the area where also the ancestral species of *Spatholobus* occurred; this species spread out over the whole area in a period when the sea level was low and the land emergent: during other periods these parts were usually inundated (= Sunda plateau) (Fig. 4.16). On the continent *S. parviflorus* (or rather its ancestor) had already developed.

For the *S. ferrugineus*-clade, the first split in the cladogram after *S. parviflorus*, it can be concluded that its ancestor speciated in a period of isolation (high sea level) from the continent of SE Asia, spreading out over the whole area of the Malay Peninsula, Sumatra, W Java and Borneo (the oldest parts of Borneo) (Fig. 4.17). At the margins of its distribution area species developed, perhaps during a later period of isolation by high sea level that isolated Borneo from the Malay Peninsula and Sumatra (Late Oligocene?): firstly *S. gyrocarpus* in the Malay Peninsula and N Sumatra; secondly *S. sanguineus* more to the north, on all of the newly uplifted parts of northern Borneo (Fig. 4.18). During later periods of lower sea level, the whole area was inhabited by the predecessor of *S. ferrugineus*, an adaptable species that was able to occupy larger regions at all latitudes, and frequently dispersed to parts outside the original range of its ancestor. In Borneo, at lower altitudes, *S. persicinus* developed, and at the same time or later on the Island of Penang *S. auritus* (Fig. 4.19). On Borneo, but only in the north, a small-flowered species developed, later splitting into two species due to the separation by a mountainous area: *S. auricomus* and *S. multiflorus*.

In the meantime speciation events occurred on the SE Asian continent as well (Fig. 4.19). The first split on the continent is that of *S. pottingeri.*, immediately after the splitting off of the *S. ferrugineus*-clade. *Spatholobus pottingeri*, developed at the margin of the total area of *Spatholobus*, became adapted to higher altitudes and is found in

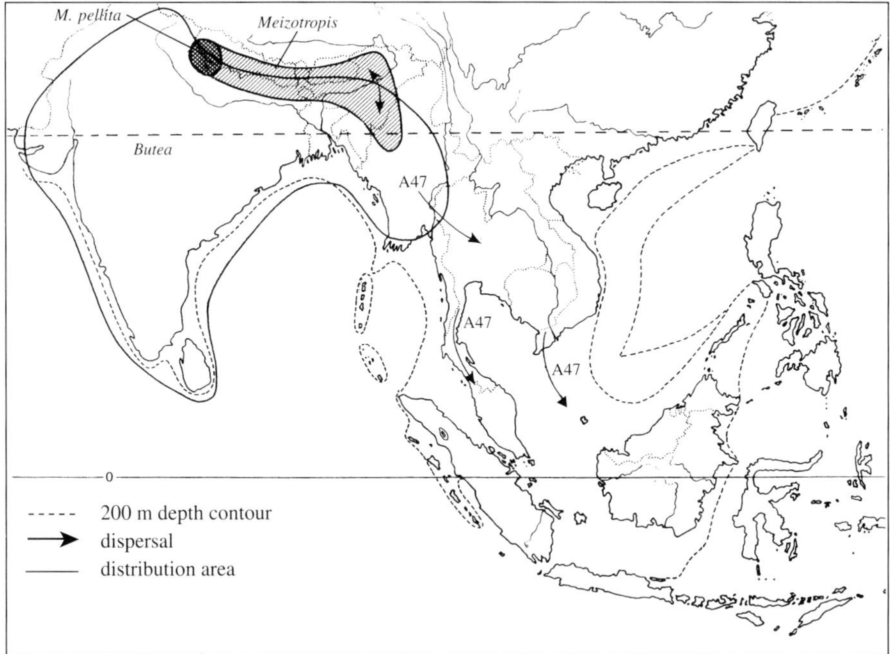

Figure 4.16. Hypothetical distributions of the ancestors of *Butea, Meizotropis* and *Spatholobus*. Dispersal of ancestor 47 (*Spatholobus*) into the Malay Archipelago.

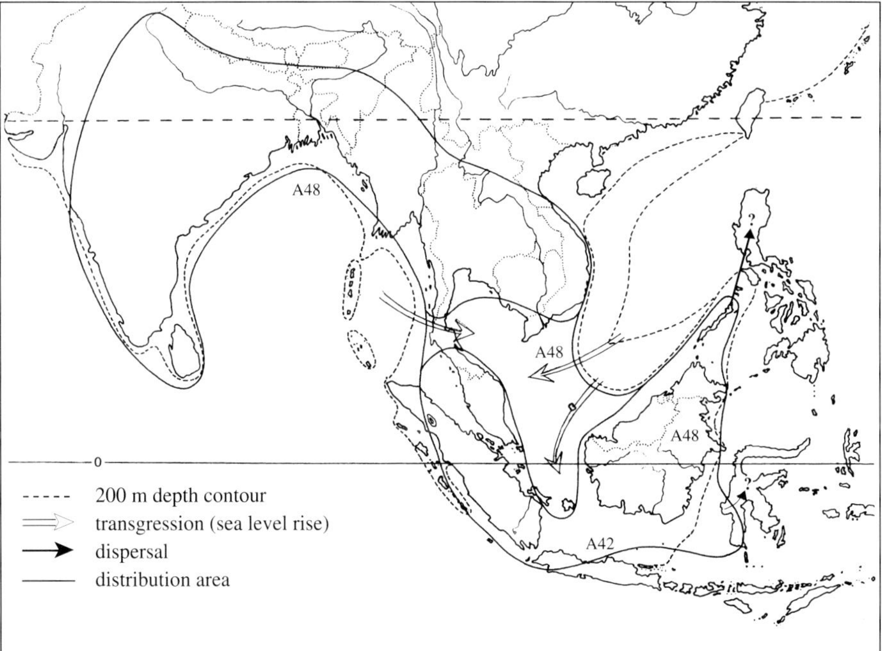

Figure 4.17. After a sea level rise part of the ancestor 48 became isolated and developed into the ancestor of the *Spatholobus ferrugineus*-clade (A42).

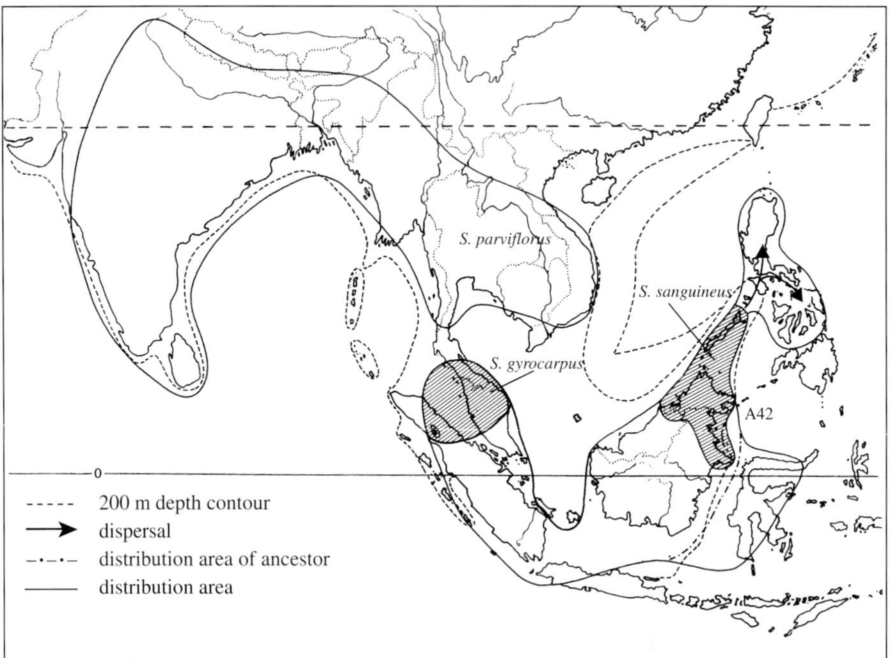

Figure 4.18. In isolated parts of the area *Spatholobus gyrocarpus* and *S. sanguineus* are speciated.

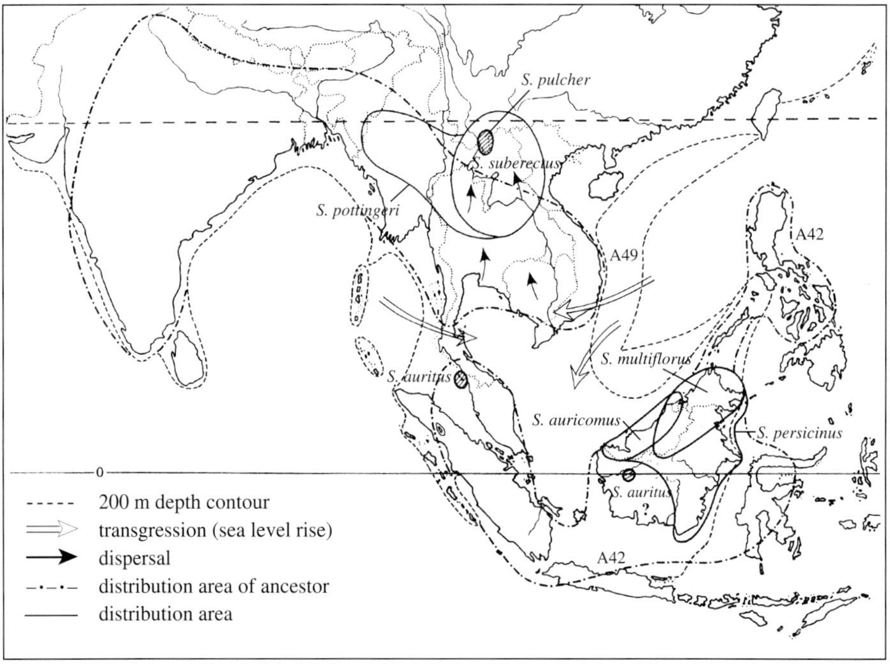

Figure 4.19. Further speciation in the *Spatholobus ferrugineus*-clade, and speciation on the continent.

the lower mountains of Yunnan / N Thailand and W Burma / E India. Once again, a speciation event on the continent has occurred, probably at the same time as the Sunda plateau became inundated (Pliocene). The climate on the continent was during this time more humid, and a tropical montane forest was present in the northern parts as well. The ancestor of the rest of *Spatholobus* (A50) probably dispersed further north, and in response to lower temperatures and a drier climate, two species developed after each other in the higher mountains of Yunnan / N Thailand, *S. pulcher* and *S. suberectus*.

The rest of *Spatholobus* (A54) remained present in the whole area, implying the option of renewed invasions from the continent into the West Malesian archipelago (Fig. 4.20). The next clade was present in the whole Sundaland area, but parts of it became isolated during high sea level (probably in the Pliocene), when not only Borneo, Sumatra and the Malay Peninsula became isolated, but the Meratus area as well. In the area in between these areas the high sea level caused large parts to inundate, and the ancestor in this area only survived in the Malay Peninsula / Sumatra and Borneo (without Meratus) and Palawan. Later, in the lower parts of the Malay Peninsula (Singapore area) and SE Sumatra, *S. ridleyi* became a separate species, and in the lower parts of Borneo *S. hirsutus* also had speciated. The remaining part of the ancestor differentiated into *S. macropterus*. Later another species, *S. oblongifolius*, managed to split off and stay in the northern part of Borneo.

After a period of low sea level another intrusion into the Sundaland area was possible. Ancestor 36 developed south of the Isthmus of Kra (Fig. 4.21). The first species that split off here is *S. littoralis*, with a rather restricted area, probably in drier parts. The second, *S. maingayi*, had a larger area, probably the same as its ancestor. The other two species differentiated in part of this area: *S. latistipulus* in SW Borneo, NE Borneo, CE Borneo, and *S. apoensis* disjunct on the Semitau and the southern Philippines.

Again, a period of relatively low sea level caused a spreading of species over a larger area. Here the same has happened as earlier in the *S. macropterus*-clade. North of the Isthmus of Kra, *S. harmandii* developed into a species adapted to drier circumstances, while south, in Penang, *S. dubius* remained as a relict of a once larger distribution, which was probably disrupted by a rising sea level (Fig. 4.22). On the northern part of Borneo two species are present, one at lower altitudes (*S. viridis*), the other in a more restricted area at higher altitudes (*S. latibractea*).

Ancestor 57 occurred in the whole SE Asian area and the northern and western parts of Borneo. Here a split between the small clade of *S. acuminatus* and the equally small rest of *Spatholobus* (A62) occurred (Fig. 4.23). The clade of *S. acuminatus* consists of only two species: *S. acuminatus* on the continent, and also on the Andaman Islands (dispersal), and *S. albus*, which occurs only on the northern part of Borneo. The ancestor of the *S. acuminatus*-clade (A64) invaded Borneo in a late stage. After an isolation event (high sea level) *S. albus* speciated on Borneo, but this species does not pass the mountain barrier of Central Borneo.

The rest of *Spatholobus* (A62) eventually became divided into three species: *S. merguensis* split off in the Mergui area; *S. purpureus* occurred later in another remote part of the area, in Kerala; the rest speciated as *S. crassifolius* in all not too northern parts (Fig. 4.23).

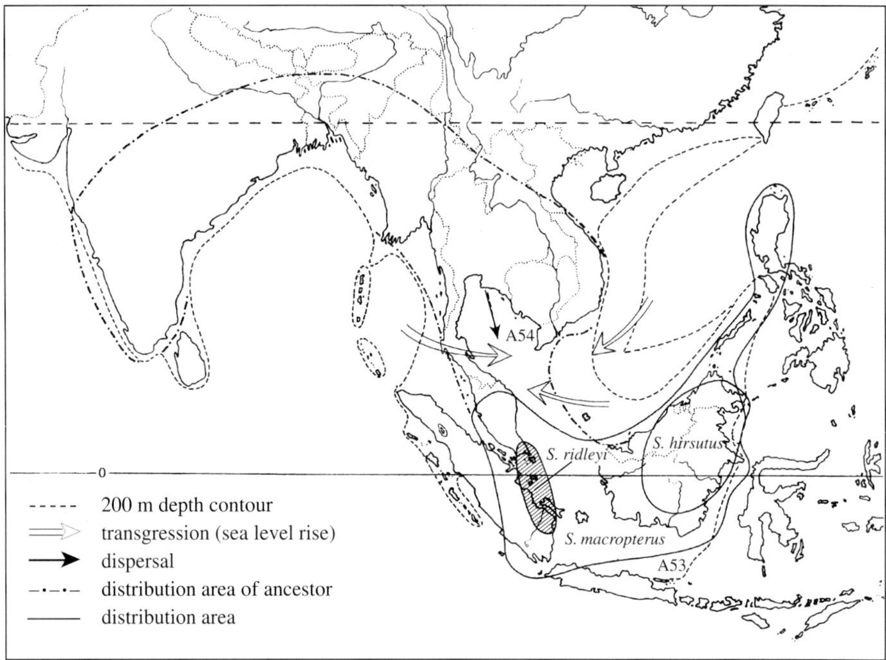

Figure 4.20. After invasion of ancestor 54 into the Malay Archipelago, once again a sea level rise. In the isolated part development of a new clade (A53).

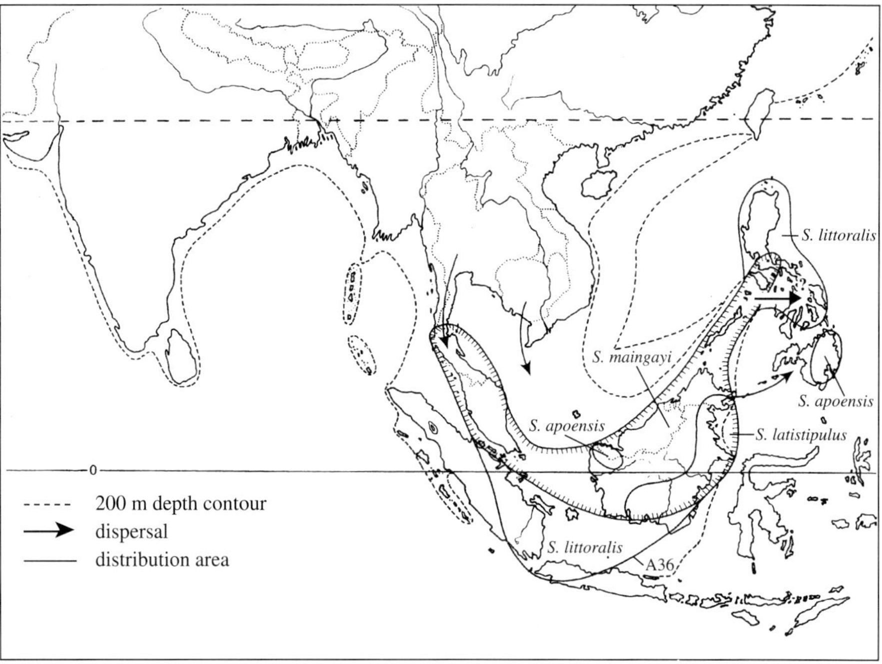

Figure 4.21. Development of the *Spatholobus maingayi*-clade after another intrusion into the Malay Archipelago.

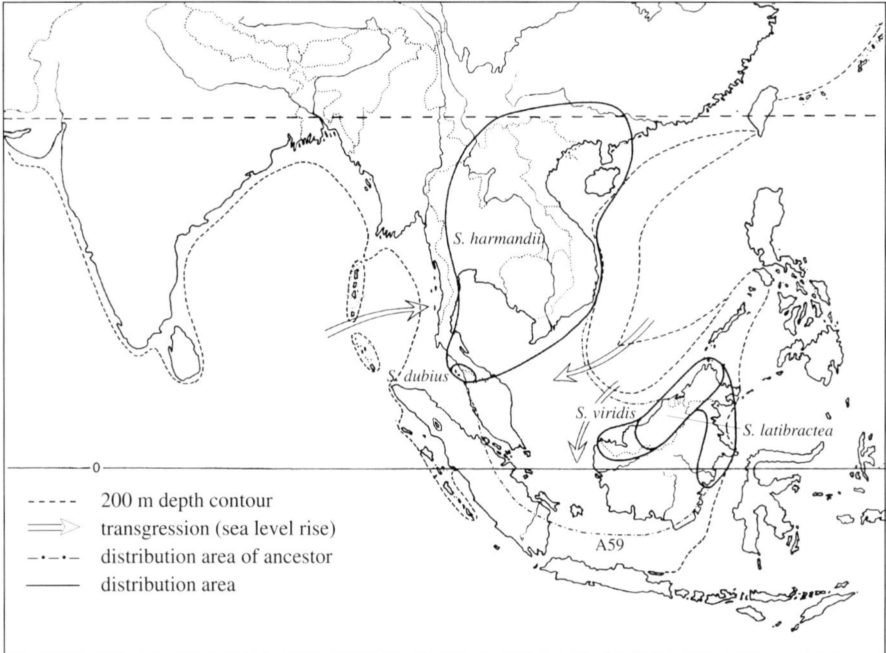

Figure 4.22. The ancestor of the *Spatholobus harmandii*-clade (A59), which spread out over the Malay Archipelago, gets isolated and speciates.

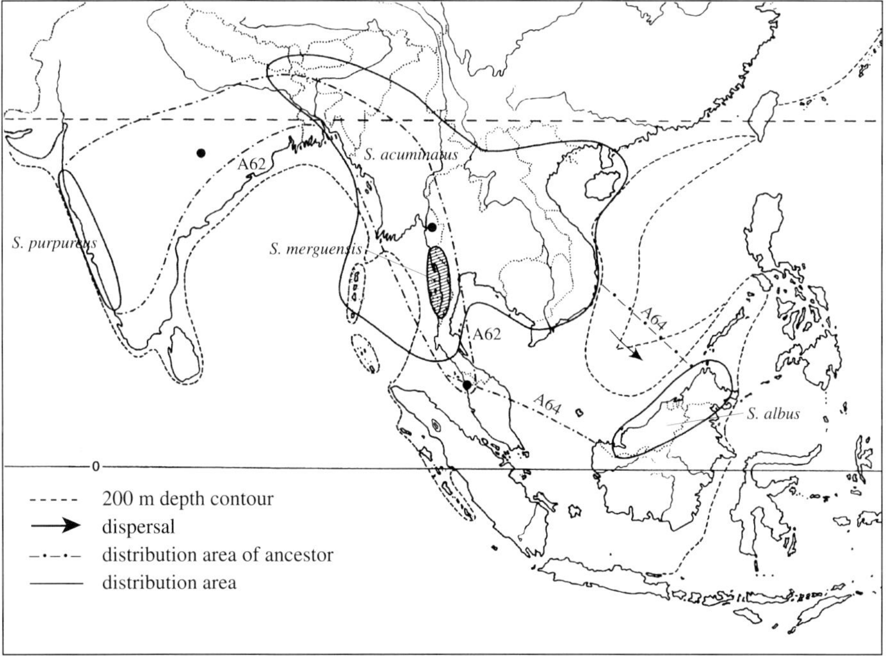

Figure 4.23. Distribution of the species and ancestors of the clades of *Spatholobus acuminatus* (A64) and *S. crassifolius* (A62).

CONCLUSIONS

Evident from the history of *Spatholobus* is that there have been many events that have led to isolation of parts of the Sundaland plateau from the Asian continent. The most obvious events are the changes in sea level, which occurred during all periods.

It would be useful if more was known about these sea level fluctuations in the SE Asian region. Studies like that of Van der Kaars and Dam (1994), describing a rather recent but long period (135,000 years) on Java, are helpful.

Spatholobus probably originated on the continent of SE Asia, together with the genera *Butea* and *Meizotropis*. After several invasions of *Spatholobus* into the West Malesian region, and after an equal number of isolation events, the present distribution pattern arose. In this history, vicariance and dispersal events both played a role. The genera *Butea* and *Meizotropis*, and the first species of *Spatholobus, S. parviflorus*, remained on the continent. The ancestor of the rest of *Spatholobus* invaded the West Malesian archipelago, after which the history of invasion and isolation began.

As far as the biogeographical analysis is concerned: this study shows that it is preferable to use more than one method of analysis. To avoid clades that are supported by only absences, it is better to use CAFCA. However, in order to make sure that all extinction events and primitive absences are also incorporated PAUP (or other parsimony procedures) is a better option. Checking the results of PAUP analyses on clades formed on absence of areas is very easy with the help of MacClade (Maddison & Maddison, 1992).

It is not enough that an area contains an endemic species, the endemic should have relations with taxa in 'related' areas as well. If there are not enough relations, this species with its distribution area – although endemic – will stand on its own and will not be very informative. Similarly it is not necessary to use only distribution areas that are areas of endemism; areas containing overlapping widespread species can be informative as well. It all depends on how informative their relatives are.

Another problem is timing. In a large area it is very well possible that more than one species develops at the same time out of the same (widespread) ancestor. For example, an ancestor occurring in the entire area from Indochina to Borneo, may develop species in each of the peripheral parts of the area: e.g., the north of Indochina and the northern part of Borneo. If this happens to be a sympatric speciation pattern, the species continuing the line of the ancestor may be present in the same area as well, or after some time may re-invade the area. This last species may in turn again be split up in other species. In this way it is evident that we recognise a split in the phylogenetic cladogram, but not in the area cladogram. It is possible that two species are placed as most related to each other because they are in fact the first split-offs from the ancestor; these two species are present in areas only related to each other by the ancestor distribution. In the N Borneo example this area is placed as sister to N Indochina, and these two together are placed as a sister group to the area in which the ancestor continued to exist.

Some areas are not very informative due to a lack of species in the area. It would be worthwhile to expand the analysis with groups occurring in the same region, but with more representatives in India, E India and W Burma, Sumatra, Sulawesi, Palawan and the Philippines. In an expanded analysis, it is expected that more information can be obtained on these areas. The groups added should preferably be of about the same geological age, and in this case not older than the Eocene. It is possible that more ancient groups will reflect a different pattern; the results will be mixed and end up in an uninformative area cladogram. Some groups that can be added are the orchid genera *Ania*, *Tainia* and *Mischobulbum* (Turner, 1992a, b), the leguminous genus *Kunstleria* (Ridder-Numan & Kornet, 1994) that has been used as an outgroup in the phylogenetic analysis in the present study, and the genus *Exacum* (Gentianaceae; Klackenberg, 1985).

On the other hand it would be worthwhile to extend the analysis with biogeographical runs on larger basic areas, in that way avoiding too many 'absent areas'. In a way the outcome will be expected to be the same, but by different optimisations it is possible that different cladograms will turn out to be the most parsimonious ones.

Another option worthwhile to take into account is the possibility to use the geological entities as found in the geological literature instead of the areas of distribution. With these geological entities it may be possible to say something about the relationships of the areas of distribution if the phylogenetic relationships of the species are reflected in their distribution. If other factors are dominant in the distribution pattern and disturb the historical pattern it will not be possible to make sensible postulations on these relationships. In this case it will be even more difficult to score the presence/ absence of a species, because the limitations are in this case not dependent on the distribution of the species. In some cases they may occur to some extent outside the chosen area. It may then be necessary to set a limit to the percentage of occurrences outside the scored area, in that way correcting for an occasional dispersal event.

I believe, however, that most will be gained by adding more genera to the analysis, thus probably giving a better supported generalised area cladogram and distribution areas that are better delimited.

REFERENCES

(to Chapters 1–4)

Adams, J.M. 1995. Influence of terrestrial ecosystems on glacial-interglacial changes in the carbon cycle. PhD thesis, University of Aix-Marseille II.

Adema, F.A.C.B. 1991. *Cupaniopsis* Radlk. (Sapindaceae): a monograph. Leiden Bot. Ser. 15: 1–190.

Audley-Charles, M.G. 1987. Dispersal of Gondwanaland: relevance to evolution of the angiosperms. In: T.C. Whitmore (ed.), Biogeographical evolution of the Malay Archipelago: 5–25. Clarendon Press, Oxford.

Audley-Charles, M.G., P.D. Ballantyne & R. Hall. 1988. Mesozoic-Cenozoic rift-drift sequence of Asian fragments from Gondwanaland. In: C.R. Scotese & W.W. Sager (eds.), Mesozoic and Cenozoic plate reconstructions. Tectonophysics 155: 317–330.

Awasthi, N. 1992. Indian fossil Legumes. In: P.S. Herendeen & D.L. Dilcher (eds.), Advances in Legume Systematics, part 4. The Fossil Record: 225–250. Royal Botanic Gardens, Kew.

Axelius, B. 1990. The genus *Xanthophytum* (Rubiaceae). Taxonomy, phylogeny and biogeography. Blumea 34: 425–497.

Axelius, B. 1991. Areas of distribution and areas of endemism. Cladistics 7: 197–199.

Axelrod, D.I. 1992. Climatic pulses, a major factor in Legume evolution. In: P.S. Herendeen & D.L. Dilcher (eds.), Advances in Legume Systematics, part 4. The Fossil Record: 259–279. Royal Botanic Gardens, Kew.

Baker, J.G. 1879. Leguminosae. In: J.D. Hooker, The Flora of British India 2: 193–194. L. Reeve, London.

Barber, A.J. & R. Hall. 1995. Tectonic evolution of Southeast Asia. Book I (lectures by A.J. Barber) and Book II (lectures by R. Hall), held at the Institute of Earth Sciences, Vrije Universiteit, Amsterdam.

Bergman, S., D.Q. Coffield, J.P. Talbot & R.A. Garrard. 1996. Tertiary tectonic and magmatic evolution of western Sulawesi and the Makassar Strait, Indonesia: evidence for a Miocene continent-continent collision. In: R. Hall & D.J. Blundell (eds.), Tectonic evolution of Southeast Asia. Geological Society of London Special Publications 106: 391–429.

Blatter, E. 1929. Revision of the genus *Butea* Koen. J. Ind. Bot. Soc. 8: 133–138.

Bremer, K. 1992. Ancestral areas: a cladistic reinterpretation of the centre of origin concept. Syst. Biol. 41: 436–445.

Bremer, K. 1995. Ancestral areas: optimization and probability. Syst. Biol. 44: 255–259.

Briggs, J.C. 1989. The historic biogeography of India: isolation or contact? Syst. Zool. 38: 322–332.

Brooks, D.R. 1990. Parsimony analysis in historical biogeography and coevolution: Methodological and theoretical update. Syst. Zool. 39: 14–30.

Brooks, D.R. & D.A. McLennan. 1991. Phylogeny, ecology, and behavior. University of Chicago Press, Chicago.

Bruneau, A., J.J. Doyle & J.L. Doyle. 1995. Phylogenetic relationships in Phaseoleae: evidence from chloroplast DNA restriction site characters. In: M. Crisp & J.J. Doyle (eds.), Advances in Legume Systematics, part 7. Phylogeny: 309–330. Royal Botanic Gardens, Kew.

Burrett, C., N. Duhig, R. Berry & R. Varne. 1991. Asian and South-western Pacific continental terranes derived from Gondwana, and their biogeographic significance. Austral. Syst. Bot. 4: 13–24.

Butterlin, J., B. Vrielynck, G. Bignot, J. Clermonte, M. Colchen, J. Dercourt, R. Guiraud, A. Poisson & L.E. Ricou. 1993. Lutetian (46–40 Ma). In: J. Dercourt, L.E. Ricou & B. Vrielynck (eds.), Atlas Tethys Palaeoenvironmental Maps: 211–223. Gauthier-Villars, Paris.

Camoin, G., Y. Bellion, J. Dercourt, R. Guiraud, J. Lucas, A. Poisson, L.E. Ricou & B. Vrielynck. 1993. Late Maastrichtian (69.5–65 Ma). In: J. Dercourt, L.E. Ricou & B. Vrielynck (eds.), Atlas Tethys Palaeoenvironmental Maps: 179–196. Gauthier-Villars, Paris.

Cavelier, C., J. Butterlin, J. Clermonte, M. Colchen, J. Dercourt, R. Guiraud, C. Lorenz & L.E. Ricou. 1993. Late Burdigalian (18–16.5 Ma). In: J. Dercourt, L.E. Ricou & B. Vrielynck (eds.), Atlas Tethys Palaeoenvironmental Maps: 225–242. Gauthier-Villars, Paris.

Cracraft, J. 1983. Cladistic analysis and vicariance biogeography. Amer. Sci. 71: 273–281.

Cracraft, J. 1988. Deep-history biogeography: Retrieving the historical patterns of evolving continental biotas. Syst. Zool. 37: 221–236.

Daly, M.C., M.A. Cooper, I. Wilson, D.G. Smith & B.G.D. Hooper. 1991. Cenozoic plate tectonics and basin evolution in Indonesia. Marine and Petroleum Geology 8: 2–21.

De Boer, A.J. 1995. Islands and cicadas adrift in the west-Pacific, biogeographic patterns related to plate tectonics. Tijdschrift voor Entomologie 138: 169–244.

De Jong, R. 1987. Biogeography – what's the question? In: P. Hovenkamp (ed.), Systematics and evolution: a matter of diversity: 251–262. Utrecht.

De Jong, R. & C.G. Treadaway. 1994. The Hesperiidae (Lepidoptera) of the Philippines. Zool. Verhand. 288: 81–115.

Dercourt, J., L.E. Ricou & B. Vrielynck (eds.). 1993. Atlas Tethys Palaeoenvironmental Maps. Gauthier-Villars, Paris, 307 pp., 14 maps, 1 plate.

Farris, J.S. 1988. Hennig86, version 1.5. Computer program and manual. University of Stony Brook, New York.

Ferguson, D.F. 1993. The impact of Late Cenozoic environmental changes in East Asia on the distribution of terrestrial plants and animals. In: N.G. Jablonski (ed.), Evolving landscapes and evolving biotas of East Asia since the Mid-Tertiary. Proc. Third Conf. on Evolution of the East Asian Environment: 145–196. Hong Kong.

Ferguson, I.K. 1984. Pollen morphology and systematics of the subfamily Papilionoideae (Leguminosae). In: W.F. Grant (ed.), Plant Biosystematics: 377–394. Academic Press, Toronto.

Ferguson, I.K. & J.J. Skvarla. 1981. The pollen morphology of the subfamily Papilionoideae (Leguminosae). In: R.M. Polhill & P.H. Raven (eds.), Advances in Legume Systematics, part 2: 859–896. Royal Botanic Gardens, Kew.

Forey, P.L., Humphries, C.J., Kitching, I.L., Scotland, R.W., Siebert, D.J. & D.M. Williams. 1993. Cladistics. A practical course in systematics. Clarendon Press, Oxford.

Geesink, R. 1981. Tephrosieae. In: R.M. Polhill & P.H. Raven (eds.), Advances in Legume Systematics, part 1: 245–260. Royal Botanic Garden, Kew.

Geesink, R. 1984. Scala Millettiearum. A survey of the genera of the tribe Millettieae (Legum.-Pap.) with methodological considerations. Leiden Bot. Series 8, Leiden University Press, Leiden.

Goloboff, P.A. 1993. Pee-Wee, version 2.0. Computer program and manual. Published by the author, New York.

Guo Shuang-Xing. 1993. The evolution of the Cenozoic tropical monsoon climate and monsoon forests in southwestern China. In: N.G. Jablonski (ed.), Evolving landscapes and evolving biotas of East Asia since the Mid-Tertiary. Proc. Third Conference on Evolution of the East Asian Environment: 123–135. Hong Kong.

Guo Shuang-Xing & Zhou Zhe-Kun. 1992. The Megafossil Legumes from China. In: P.S. Herendeen & D.L. Dilcher (eds.), Advances in Legume Systematics, part 4. The Fossil Record: 207–223. Royal Botanic Gardens, Kew.

Hall, R. 1995. Plate tectonic reconstructions of the Indonesian region. Indonesian Petroleum Association Proc. 24th Annual Convention 1995.

Hall, R. 1996. Reconstructing Cenozoic SE Asia. In: R. Hall & D.J. Blundell (eds.), Tectonic evolution of Southeast Asia. Geological Society of London Special Publications 106: 153–184.

Hall, R. & A.J. Barber. 1995. Tectonic evolution of Southeast Asia. Book I (lectures by A.J. Barber) and Book II (lectures by R. Hall), held at the Institute of Earth Sciences, Vrije Universiteit, Amsterdam.

Hamilton, W.B. 1973. Tectonics of the Indonesian region. Geol. Soc. Malaysia Bull. 6: 3–10.

Hamilton, W.B. 1979. Tectonics of the Indonesian region. U.S. Geological Survey Professional Paper 1078.

Hamilton, W.B. 1988. Plate tectonics and island arcs. Geological Soc. of America Bull. 100: 1503–1527.

Herendeen, P.S., W.L. Crepet & D.L. Dilcher. 1992. The fossil history of the Leguminosae: phylogenetic and biogeographic implications. In: P.S. Herendeen & D.L. Dilcher (ed.), Advances in Legume Systematics, part 4. The Fossil Record: 303–316. Royal Botanic Gardens, Kew.

Hutchison, C.S. 1989a. Geological evolution of South-East Asia. Oxford Monogr. on Geology and Geophysics, no.13. Clarendon Press, Oxford.

Hutchison, C.S. 1989b. The Palaeo-Tethyan Realm and Indosinian Orogenic System of Southeast Asia. In: A.M.C. Sengör (ed.), Tectonic evolution of the Tethyan Region. Kluwer Academic Publishers, Dordrecht.

Hutchison, C.S. 1992. The Eocene unconformity on Southeast and East Sundaland. Geol. Soc. Malaysia Bull. 32: 69–88.

Hutchison, C.S. 1996. The 'Rajang accretionary prism' and 'Lupar Line' problem of Borneo. In: R. Hall & D.J. Blundell (eds.), Tectonic evolution of Southeast Asia. Geological Society of London Special Publications 106: 247–262.

Klackenberg, J. 1985. The genus *Exacum* (Gentianaceae). Opera Botanica 84: 1–144.

Klackenberg, J. 1995. Taxonomy and phylogeny of the SE Asian genus *Genianthus* (Asclepiadaceae). Bot. Jahrb. 117: 401–467.

Lackey, J.A. 1981. Phaseoleae DC. (1825). In: R.M. Polhill & P.H. Raven (eds.), Advances in Legume Systematics, part 1: 301–327. Royal Botanic Gardens, Kew.

Lam, H.J. 1930. Het genetisch-plantengeografisch onderzoek van den Indischen Archipel en Wegener's verschuivingstheorie. Tijdschr. Kon. Ned. Aardrijkskundig Genootschap 47 (4): 553–581.

Lavin, M. 1995. Tribe Robinieae and allies; model groups for assessing early Tertiary northern latitude diversification of tropical legumes. In: M. Crisp & J.J. Doyle (eds.), Advances in Legume Systematics, part 7. Phylogeny: 141–160. Royal Botanic Gardens, Kew.

Lavin, M. & M. Luckow. 1993. Origins and relationships of tropical North America in the context of the boreotropics hypothesis. Amer. J. Bot. 80: 1–14.

Lee, T.-Y. & L.A. Lawver. 1992. Cenozoic plate reconstruction of the South China Sea Region. Abstracts of the 29th International Geological Congress, Kyoto, Japan, Aug./Sept., 1992. In: Journal of Tertiary / Quaternary Research in Southeast Asia.

Lorenz, C., J. Butterlin, C. Cavelier, J. Clermonte, M. Colchen, J. Dercourt, R. Guiraud, C. Montenat, A. Poisson, L.E. Ricou & M. Sandulescu. 1993. Late Rupelien (30–28 Ma). In: J. Dercourt, L.E. Ricou & B. Vrielynck (eds.), Atlas Tethys Palaeoenvironmental Maps: 211–224. Gauthier-Villars, Paris.

Maddison, W.P. & D.R. Maddison. 1992. MacClade: Analysis of phylogeny and character evolution. Version 3.0. Sinauer Associates, Sunderland, Massachusetts.

Masse, J.P., Y. Bellion, J. Benkhelil, J. Dercourt, R. Guiraud & L.E. Ricou. 1993. Lower Aptian (114–112 Ma). In: J. Dercourt, L.E. Ricou & B. Vrielynck (eds.), Atlas Tethys Palaeoenvironmental Maps: 135–152. Gauthier-Villars, Paris.

McCourt, W.J., M.J. Crow, E.J. Cobbing & T.C. Amin. 1996. Mesozoic and Cenozoic plutonic evolution of SE Asia: evidence from Sumatra, Indonesia. In: R. Hall & D.J. Blundell (eds.), Tectonic evolution of Southeast Asia. Geological Society of London Special Publications 106: 321–335.

Metcalfe, C.R. & L. Chalk. 1979. Anatomy of the Dicotyledons 1. Clarendon Press, Oxford.

Metcalfe, I. 1988. Origin and assembly of South-East Asian continental terranes. In: M.G. Audley-Charles & A. Hallam (eds.), Gondwana and Tethys. Geological Society of London Special Publications 37: 101–118.

Metcalfe, I. 1990. Allochthonous terrane processes in Southeast Asia. Philos. Trans. Roy. Soc. London A 331: 625–640.

Metcalfe, I. 1991. Late Palaeozoic and Mesozoic palaeogeography of Southeast Asia. Palaeogeogr., Palaeoclimatol., Palaeoecol. 87: 211–221.

Metcalfe, I. 1994a. Gondwanaland origin, dispersion, and accretion of East and Southeast Asian continental terranes. J. South Amer. Earth Sciences 7: 333–347.

Metcalfe, I. 1994b. Late Palaeozoic and Mesozoic palaeogeography of eastern Pangea and Tethys. In: A.F. Embry, B. Beauchamp & D.J. Glass (eds.), Pangea: Global environments and resources. Can. Soc. Petrl. Geol., Memoir 17: 97–111.

Metcalfe, I. 1996. Pre-Cretaceous evolution of SE Asian terranes. In: R. Hall & D.J. Blundell (eds.), Tectonic evolution of Southeast Asia. Geological Society of London Special Publications 106: 97–122.

Moog, J. & U. Maschwitz. 1994. Associations of *Cladomyrma* (Hym., Formicidae, Formicinae) with plants in SE Asia. In: A. Lenoir, G. Arnold & M. LePage (eds.), Les Insectes Sociaux: 173. 12th Congr. Intern. Union for Study of Social Insects IUSSI, Paris, 21–27 Aug. 1994.

Morley, R.J. & J.R. Flenley. 1987. Late Cenozoic vegetational and environmental changes in the Malay Archipelago. In: T.C. Whitmore (ed.), Biogeographical evolution of the Malay Archipelago: 50–59. Clarendon Press, Oxford.

Morrone, J.J. & J.M. Carpenter. 1994. In search of a method for cladistic biogeography: an empirical comparison of Component analysis, Brooks Parsimony analysis, and Three-Area Statements. Cladistics 10: 99–153.

Nelson, G. & P. Ladiges. 1991. Standard assumptions for biogeographic analysis. Austral. Syst. Bot. 4: 41–58.

Nelson, G. & P. Ladiges. 1992. TAS (MS-DOS computer program). Published by the authors, New York and Melbourne.

Nelson, G. & N.I. Platnick. 1981. Systematics and biogeography: cladistics and vicariance. Columbia University Press, New York.

Page, R.D.M. 1988. Quantitative cladistic biogeography: constructing and comparing area cladograms. Syst. Zool. 37: 254–270.

Page, R.D.M. 1990. Component analysis: a valiant failure? Cladistics 6: 119–136.

Page, R.D.M. 1993. COMPONENT, version 2.0. Computer program and manual. Natural History Museum, London.

Patterson, C. & H.G. Owen. 1991. Indian isolation or contact? A response to Briggs. Syst. Zool. 40: 96–100.

Pigram, C.L. & H.L. Davies. 1987. Terranes and the accretion history of the New Guinea orogen. BMR J. Austral. Geology & Geophysics 10: 193–211.

Potts, R. & A.K. Behrensmeyer. 1992. Late Cenozoic terrestrial ecosystems. In: A.K. Behrensmeyer et al. (eds.), Terrestrial ecosystems through time. Evolutionary Paleoecology of terrestrial plants and animals: 397–515. University of Chicago Press, Chicago, London.

Rangin, C., L. Jolivet, M. Pubellier & the Tethys Pacific working group. 1990a. A simple model for the tectonic evolution of Southeast Asia and Indonesian region for the past 43 m.y. Bull. Soc. géol. France, 8, t. 6: 889–905.

Rangin, C., M. Pubellier, J.Azéma, A. Briais, P.Chotin, H.Fontaine, P.Huchon, L. Jolivet, R. Maury, C. Muller, J.-P. Rampnoux, J.-F. Stephan, J. Tournon, N. Cottereau, J. Dercourt & L.-E. Ricou. 1990b. The quest for Tethys in the western Pacific. 8 paleogeodynamic maps for Cenozoic time. Bull. Soc. géol. France, 8, t. 6: 909–913 + 8 colour maps.

Raven, P.H. & R.M. Polhill. 1981. Biogeography of the Leguminosae. In: R.M. Polhill & P.H. Raven (eds.), Advances in Legume Systematics, part 1: 27–34. Royal Botanic Gardens, Kew.

Ridder-Numan, J.W.A. 1992. *Spatholobus* (Leguminosae–Papilionoideae): a new species and some notes. Blumea 37: 63–71.

Ridder-Numan, J.W.A. 1995. Phylogeny and biogeography of *Spatholobus*, *Butea*, *Meizotropis* and *Kunstleria* (Leguminosae–Papilionoideae). In: M. Crisp & J.J. Doyle (eds.), Advances in Legume Systematics, part 7. Phylogeny: 133–139. Royal Botanic Gardens, Kew.

Ridder-Numan, J.W.A. & R.W.J.M. van der Ham (Submitted and accepted). Pollen morphology of *Butea, Kunstleria, Meizotropis* and *Spatholobus* (Leguminosae–Papilionoideae), with notes on their position in the tribes Millettieae and Phaseoleae. Review of Palynology and Paleobotany. — This Thesis: Chapter 5.

Ridder-Numan, J.W.A. & D.J. Kornet. 1994. A revision of the genus *Kunstleria* (Leguminosae–Papilionoideae). Blumea 38: 465–485.

Ridder-Numan, J.W.A. & H. Wiriadinata. 1985. A revision of the genus *Spatholobus* (Leguminosae–Papilionoideae). Reinwardtia 10: 139–205.

Sanderson, M.J. & M.J. Donoghue. 1989. Patterns of variation in levels of homoplasy. Evolution 43: 1781–1795.

Sanjappa, M. 1987. Revision of the genera *Butea* Roxb. ex Willd. and *Meizotropis* Voigt (Fabaceae). Bull. Bot. Surv. India 29: 199–225.

Schot, A.M. 1991. Phylogenetic relations and historical biogeography of *Fordia* and *Imbralyx* (Papilionaceae: Millettieae). Blumea 36: 205–234.

Simandjuntak, T.O. & A.J. Barber. 1996. Contrasting tectonic styles in the Neogene orogenic belts of Indonesia. In: R. Hall & D.J. Blundell (eds.), Tectonic evolution of Southeast Asia. Geological Society of London Special Publications 106: 185–201.

Sosef, M.S.M. 1994. Refuge begonias. Taxonomy, phylogeny and historical biogeography of *Begonia* sect. *Loasibegonia* and sect. *Scutobegonia* in relation to glacial rain forest refuges in Africa. Wageningen Agric. Univ. Papers 94-1: 1–306.

Swofford, D.L. 1991. PAUP: Phylogenetic analysis using parsimony, version 3.0s. Computer program and manual. Illinois Natural History Survey, Champaign.

Swofford, D.L. 1993. PAUP: Phylogenetic analysis using parsimony, version 3.1.1. Computer program and manual. Illinois Natural History Survey, Champaign.

Taylor, B. & D.E. Hayes. 1983. Origin and history of the South China Sea Basin. In: D.E. Hayes (ed.), The tectonic and geologic evolution of South East Asian Seas and Islands, part 2: 23–56. Geophysical Monographs, Washington.

Thewissen, J.G.M. & M.C. McKenna. 1992. Paleobiogeography of Indo-Pakistan: a response to Briggs, Patterson and Owen. Syst. Biol. 41: 248–251.

Turner, H. 1992a. A revision of the orchid genera *Ania, Hancockia, Mischobulbum*, and *Tainia*. Orchid Monographs 6: 43–100. Rijksherbarium / Hortus Botanicus, Leiden.

Turner, H. 1992b. Revision and cladistics of *Ania, Tainia, Mischobulbum*, and *Hancockia* (Orchidaceae). Acta Bot. Neerl. 41: 108–109.

Turner, H. 1995. Cladistic and biogeographic analysis of *Arytera* Blume and *Mischarytera* gen. nov. (Sapindaceae), with notes on methodology and a full taxonomic revision. Blumea Suppl. 9: 1–230. Rijksherbarium / Hortus Botanicus, Leiden.

Turner, H. & M. Zandee.1995. The behaviour of Goloboff's tree fitness measure F. Cladistics 11: 57–72.

Van der Kaars, W.A. & M.A.C. Dam. 1994. A 135,000-year record of vegetational and climatic change from the Bandung area, West-Java, Indonesia. Palaeogeogr., Palaeoclimatol., Palaeoecol. 117: 55–72.

Van der Werff, B. 1996. Forearc development and early orogenesis along the eastern Sunda / western Banda Arc (Indonesia). Thesis, Vrije Universiteit, Amsterdam.

Van Steenis, C.G.G.J. 1961. Introduction. In: M.S. van Meeuwen, H.P. Nooteboom & C.G.G.J. van Steenis, Preliminary revision of some genera of Malaysian Papilionaceae I. Reinwardtia 5: 420–429.

Van Steenis, C.G.G.J. 1979. Plant-geography of east Malesia. Bot. J. Linn. Soc. 79: 97–178.

Van Welzen, P.C. 1989. *Guioa* Cav. (Sapindaceae): taxonomy, phylogeny, and historical biogeography. Leiden Bot. Ser. 12: 1–315.

Van Welzen, P.C. 1992. Interpretation of historical biogeographical results. Acta Bot. Neerl. 41: 75–87.

Veevers, J.J. 1991. Phanerozoic Australia in the changing configuration of Proto-Pangea through Gondwanaland and Pangea to the present dispersed continents. Austral. Syst. Bot. 4: 1–11.

Vermeulen, J.J. 1993. A taxonomic revision of *Bulbophyllum*, sections *Adelopetalum, Lepantanthe, Macrouris, Pelma, Peltopus*, and *Uncifera* (Orchidaceae). Orchid Monographs 7: 1–324. Rijksherbarium / Hortus Botanicus, Leiden.

Verstappen, H.Th. 1975. On palaeoclimates and landform development in Malesia. In: G.-J. Bartstra & W.A. Casparie (ed.), Modern Quaternary research in Southeast Asia 1: 3–35. A.A. Balkema, Rotterdam.

Verstappen, H.Th. 1980. Quaternary climate changes and natural environment in SE Asia. GeoJournal 4.1: 45–54.

Walker, D. 1982. Speculations on the origin and evolution of Sunda–Sahul rain forests. In: G.T. Prance (ed.), Biological diversification in the tropics: 554–575. Columbia University Press, New York.

Wegener, A. 1915. Die Enstehung der Kontinente und Ozeane. Vieweg, Brunswick.

Weijermars, R. 1989. Global tectonics since the break-up of Pangea 180 million years ago: evolution maps and lithospheric budget. Earth-Science Rev. 26: 113–162.

Wensink, H. 1996. The structural history of Sundaland: Paleomagnetic implications from the late Cretaceous Laispu formations of Sumba. Submitted to Tectonophysics.

Wheeler, E.A. & P.Baas. 1992. Fossil wood of the Leguminosae: a case study in xylem evolution and ecological anatomy. In: P.S. Herendeen & D.L. Dilcher (ed.), Advances in Legume Systematics, part 4. The Fossil Record: 281–301. Royal Botanic Gardens, Kew.

Whitmore, T.C. 1981. Palaeoclimate and vegetation history. In: T.C. Whitmore (ed.), Wallace's line and plate tectonics: 36–42. Oxford.

Wiens, J.J. 1995. Polymorphic characters in phylogenetic systematics. Syst. Biol. 44: 482–500.

Wiley, E.O. 1981. Phylogenetics: the theory and practice of phylogenetic systematics. Wiley-Interscience, New York.

Wiley, E.O. 1987. Methods in vicariance biogeography. In: P. Hovenkamp et al. (eds.), Systematics and evolution: a matter of diversity: 283–306. Utrecht University.

Wing, S.L. & H.-D. Sues. 1992. Mesozoic and Early Cenozoic terrestrial ecosystems. In: A.K. Behrensmeyer et al. (eds.), Terrestrial ecosystems through time. Evolutionary Paleoecology of terrestrial plants and animals: 327–394. University of Chicago Press, Chicago, London.

Zandee, M. 1995. CAFCA – a Collection of APL Functions for Cladistic Analysis. Macintosh and PowerPC version 1.5d. Computer program and manual. Institute of Evolutionary and Ecological Sciences, Leiden University, Leiden.

Zandee, M. & M.C. Roos. 1987. Component-compatibility in historical biogeography. Cladistics 3: 305–332.

Chapter 5

POLLEN MORPHOLOGY

Pollen morphology of *Butea, Kunstleria, Meizotropis,* and *Spatholobus* (Leguminosae–Papilionoideae), with notes on their position in the tribes Millettieae and Phaseoleae

J.W. A. Ridder-Numan & R.W. J.M. van der Ham
Rijksherbarium / Hortus Botanicus, Leiden, The Netherlands

[Submitted to and accepted for Review of Palaeobotany and Palynology]

ABSTRACT

Pollen descriptions are given of *Butea, Kunstleria, Meizotropis,* and *Spatholobus* based on LM, SEM, and TEM data. The pollen of *Spatholobus* is more diverse than that of the other genera. Most species have tricolporate pollen, but (para)syncolporate pollen occurs in several *Spatholobus* species. The ornamentation is rugulate in *Kunstleria,* perforate to microreticulate in *Butea,* fossulate to verrucate in *Meizotropis,* and psilate-perforate to fossulate or microreticulate in *Spatholobus.* The infratectal layer is mostly columellate in *Kunstleria* and *Butea,* more or less granular in *Meizotropis,* and either columellate or coarsely granular in *Spatholobus.* The foot layer is well-developed in most species except for *Butea, Meizotropis,* and *Spatholobus parviflorus,* where it is very thin compared to the endexine. In *Spatholobus,* three types and three subtypes can be recognised. The pollen of *Kunstleria* closely resembles that of the tribe Millettieae. Pollen of *Spatholobus* shows features of both Millettieae and Phaseoleae pollen. Pollen of *Butea* and especially that of *Meizotropis* is similar to Phaseoleae pollen. The results of a cladistic analysis of the four genera agree with trends in ornamentation, aperture system, and infratectal structure reported for Phaseoleae pollen in the literature.

INTRODUCTION

As part of a research programme at the Rijksherbarium / Hortus Botanicus on the historical biogeography of the Malesian Archipelago, the genera *Butea* Roxb. ex Willd., *Kunstleria* Prain, *Meizotropis* Voigt and *Spatholobus* Hassk. have been studied macromorphologically, leaf anatomically and pollen morphologically, and have been subjected to a phylogenetic analysis, resulting in the present thesis. The pollen morphological part of the study is presented here as a separate chapter.

All four genera have been recently revised, *Kunstleria* by Ridder-Numan and Kornet (1994), *Spatholobus* by Ridder-Numan and Wiriadinata (1985) and Ridder-Numan (1992), *Butea* and *Meizotropis* by Sanjappa (1987). *Kunstleria* consists of eight species, seven of which occur in the Malesian region and one in Kerala (India). In the genus *Spatholobus* 29 species are recognised, all occurring in Southeast Asia. The genera *Butea* and *Meizotropis* both consist of two species occurring on the mainland Southeast Asia.

In addition to increasing a database for the phylogenetic analysis, the aim of the present study was to clarify the tribal placement of *Spatholobus*, which hitherto has not been satisfactory. The genera belong to two tribes of the Papilionoideae; *Butea* and *Meizotropis* belong to the Phaseoleae (usually in the subtribe Erythrininae), while *Kunstleria* is included in the tribe Millettieae (Geesink, 1981, 1984; Polhill, 1981; Ridder-Numan & Kornet, 1994). *Spatholobus* is traditionally placed in the Phaseoleae, subtribe Erythrininae (Lackey, 1981), although other subtribes have been suggested as well, e.g., Glycininae (Geesink, 1984). However, *Spatholobus* is ambiguous and has features of both tribes: the leaves and pods are like those in the Phaseoleae, but the rather unspecialised flowers are more like those of the Millettieae. *Spatholobus* has long been regarded as a part of *Butea* by various authors; especially *S. parviflorus*, of which foliage and pod closely resemble those of *Butea*. Based on macromorphological characters, it seems reasonable to regard *Spatholobus* as basal in the Phaseoleae, which are regarded as a more advanced tribe than the Millettieae (Lackey, 1981; Ridder-Numan & Wiriadinata, 1985). In a phylogenetic analysis of chloroplast DNA restriction site characters in the Phaseoleae (Bruneau et al., 1995), *Spatholobus* was placed next to *Butea* and closer to the Phaseoleae than to the Millettieae.

The pollen morphology of *Butea* and *Meizotropis* has been studied by Murthy (1989, 1992). Occasionally, species are described in pollen-floristic papers: *B. monosperma* in Shome et al. (1980), Mandal & Chanda (1981), Mondal & Mondal (1983), Gupta & Sharma (1986; as *B. frondosa*), Banik et al. (1986), Okolo & Gill (1986), and *B. monosperma* and *B. superba* in Vishnu-Mittre & Sharma (1962) and Tewari & Nair (1978, 1979). Graham and Spencer Tomb (1977) compare the pollen of *B. monosperma* (also as *B. frondosa*), *Spatholobus parviflorus* (as *B. parviflora* and *B. roxburghii*), and *S. harmandii* with that of *Erythrina* (Phaseoleae).

The pollen morphology of *Kunstleria* and *Spatholobus* has never been studied in detail. A short description of the pollen of *Kunstleria philippinensis* is given by Bulalacao (1990), of *Spatholobus parviflorus* (as *B. parviflora* and *S. roxburghii*) by Vishnu-Mittre & Sharma (1962) and Tewari & Nair (1978, 1979), and of *S. parviflorus* and *S. harmandii* by Graham & Spencer Tomb (1977).

The present paper describes the pollen morphology of nearly all species of *Butea*, *Meizotropis*, *Kunstleria* and *Spatholobus*; pollen of *Kunstleria curtisii* Prain, *K. keralensis* Mohanan & Nair and *Spatholobus bracteolatus* Prain ex King was not available.

MATERIAL AND METHODS

If available, 10 samples per species were taken from herbarium material, preferably anthers from mature buds. In case of dimorphous anthers, the largest ones were taken (see also the note under *Kunstleria* in Results). All samples were prepared for light microscopy (LM) and scanning electron microscopy (SEM), and a selection of the samples for transmission electron microscopy (TEM; samples indicated in the specimen list) according to the techniques used by Van der Ham (1990). If possible 10 grains per sample were measured. The thickness of the exine layers was measured in the central part of a mesocolpium (SEM and TEM). The extremes of P and E per spe-

Table 5.1. Characters and measures (in μm) of the species of *Butea*, *Kunstleria*, *Meizotropis*, and *Spatholobus*. — con = continuous, discon = discontinuous; psil = psilate, perf = perforate, fos = fossulate, microret = microreticulate, rug = rugulate, ver = verrucate; somet = sometimes; (T) = TEM values, (S) = SEM values.

Species	foot layer	ft l/endexine	infratectal structure	ornamentation	colpi fused	endoaperture	colpus width	type
B. monosperma	discon	1/8–1/4	short columellae/granule-like columellae	microret/coarsely fos	no	4–7 × 8–10	5–9	*Butea*
B. superba	?	?	short columellae/granule-like columellae	microret	no	6–8 × 4–10	10	*Butea*
K. forbesii	con	1/2–1 1/2	irregular columellae/granules	rug/fos/deviating	no	4–6 × 5–8	1–2(–5)	*Kunstleria*
K. geesinkii	?	?	irregular columellae	rug/fos	no	5–9 × 8–9	2–4.5	*Kunstleria*
K. kingii	?	?	?	rug	no	5 × 4–9	1.5–3	*Kunstleria*
K. philippinensis	?	?	?	rug-fos	no	5–6 × 7–9	2–3	*Kunstleria*
K. ridleyi	discon	1/2–1 1/3	columellae	rug/rug-fos	no	4–5 × 6–14	2–5	*Kunstleria*
K. sarawakensis	?	?	columellae	rug	no	5 × 8–9	3	*Kunstleria*
M. buteiformis	discon	1/30–1/18	indistinct/granules or granule-like columellae	coarsely fos-microret/ver/ver-fos	no	5–14 × 5–7	3–10	*Meizotropis*
M. pellita	discon	1/16–1/14	indistinct/granules or granule-like columellae	ver-fos/ver	no	5–7 × 5–7	3–5	*Meizotropis*
S. acuminatus	con	3/5–2	irregular columellae	psil-perf/fos	somet	5–10 × 8–14	4–6	IIIb
S. albus	discon	1/4–1/3	coarse granules/granule-like columellae	psil-perf	somet	?	3	IIIb
S. apoensis	discon	3/4–1 1/2	granules/irregular short columellae	microret/coarsely perf-fos/deviating	no	6–9 × 2–4	2–5	II
Elmer 11795	–	–	–	–	–	–	–	–
S. auricomus	?	?	granules/short columellae	psil-perf/fos	no	?	?	IIIb
S. auritus	?	?	indistinct/dense	psil-perf	no	5 × 14	4–6	IIIc
S. crassifolius	?	?	coarse granules	psil-perf	somet	5–6 × 7–15	5	IIIb
S. dubius	?	?	coarse granules	psil-perf	somet	4 × 8–9	1–4	IIIb
S. ferrugineus	discon	1/6–1/3	well-spaced columellae	psil-perf/(fos)	no	7 × 7	(2–)4–8(–10)	IIIa
S. gyrocarpus	?	?	?	psil-perf	no	3 × 8	4	IIIa
S. harmandii	?	?	short columellae	psil-perf/(fos)	often	5–8 × 5–6	3–5	IIIb
S. hirsutus	con	2/5–1/2	(irregular) columellae	psil-perf/fos	no	6–10 × 5–13	4–8	IIIa
S. latibractea	con	5/9–1 1/3	dense irregular columellae	psil-perf/coarsely perf	somet	6–7 × 10–12	2–3.5	IIIc
S. latistipulus	?	?	granules	fos/microret	no	?	2	II
S. littoralis	con	1/5–1 1/6	coarse granules/irregular short columellae	fos/(coarsely) perf/deviating	somet	3–5 × 7–10	1–2	II
S. macropterus	discon	1/11–2/5	well-spaced (irregular) columellae	psil-perf/fos	no	4–8 × 9–15	(1–)2–5(–10)	IIIa
S. maingayi	discon	3/5–5/4	granules	microret/fos/deviating/(coarsely) perf	somet	3–6 × 8–10	(1–)2–5	II
S. merguensis	?	?	dense granules	psil-perf	somet	8 × 6	2–3	IIIc
S. multiflorus	con	1/2	dense (irregular) columellae	psil-perf	no	5–7 × 11–15	3–5	IIIb
S. oblongifolius	con	1/6–1/3	coarse granules/granule-like columellae	psil-perf/fos/(rug)/(deviating)	somet	5–7 × 10–15	4–9	IIIb
S. parviflorus	con	1/20–1/9	granules/granule-like columellae	psil-perf/fos/(rug)	no	6–7 × 12–13	3–10	I
S. persicinus	discon	1/3–1	coarse granules/granule-like columellae	psil-perf/fos/coarsely perf	often	5–7 × 10–12	2–5(–10)	IIIb
S. pottingeri	?	?	dense granules/irregular columellae	psil-perf	no	6 × 10	3–5	IIIc
S. pulcher	?	?	?	psil-(perf)	no	?	4	IIIc
S. purpureus	discon	1/3.	granules/irregular columellae	psil-perf	somet	?	?	IIIb
S. ridleyi	?	?	?	fos	no	?	?	IIIb
S. sanguineus	con	1/4–1	(irregular) well-spaced columellae	psil-perf	no	5–7 × 10–11	2–5	IIIa
S. suberectus	con	1/6–1/3	dense coarse granules/granule-like columellae	psil-perf/fos	no	3–5–8 × 8–11	1–4	IIIc
S. viridis	discon	1/5–1/2	coarse granules/irregular columellae	psil-perf/fos	often	5–7 × 8–12	1–2(–3)	IIIb

(Table 5.1 continued)

Species	tectum	infratectum	foot layer	endexine	exine	nexine	P	E	P/E	A/E
B. monosperma	0.2-0.4 (T)	0.13-0.3 (T)	0.06-0.13 (T)	0.26-0.53 (T)	0.93-1.13 (T)	0.4-0.66 (T)	34.3 (38.4) 42	31.8 (34.7) 37.3	1.11	0.30 (0.36) 0.44
B. superba	0.33-0.55 (S)	0.33-0.44 (S)	?	?	1.33-1.44 (S)	0.4-0.8 (S)	36 (38.2) 41.2	33 (35.8) 38.3	1.07	0.36 (0.43) 0.51
K. forbesii	(0.16-)0.38-0.66 (T)	0.33-0.44 (T)	0.05-0.11 (T)	0.11-0.22 (T)	0.83-1.1 (T)	0.22-0.39 (T)	20.2 (22.8) 24.2	19.1 (20.8) 22.5	1.1	0.35 (0.39) 0.45
K. geesinkii	0.3-0.6 (S)	0.1-0.4 (S)	?	?	?	0.2-0.5 (S)	21.2 (23.5) 25.8	19.4 (20.7) 24	1.1	0.38 (0.46) 0.53
K. kingii	0.2-0.4 (S)	0.2-0.1 (S)	?	?	?	0.2-0.7 (S)	18 (20.5) 22.5	17.5 (19.8) 22	1.08	0.32 (0.4) 0.48
K. philippinensis	0.3 (S)	0.1-0.3 (S)	?	?	?	0.3-0.4 (S)	24 (26.7) 30	18 (21.4) 21	1.25	0.48 (0.52) 0.55
K. ridleyi	0.35-0.5 (T)	0.14-0.35 (T)	0.14-0.24 (T)	0.1-0.4 (T)	1.07-1.17 (T)	0.36-0.53 (T)	21.4 (23.4) 25.6	20.3 (21.8) 23.8	1.07	0.34 (0.41) 0.49
K. sarawakensis	0.2 (S)	0.2 (S)	?	?	?	0.2-0.7 (S)	25 (25.7) 27	22 (22.5) 24	1.14	0.4 (0.49) 0.58
M. buteiformis	0.13-0.46 (T)	0.06-0.13 (T)	0.03-0.06 (T)	0.67-0.7 (T)	1-1.2 (T)	0.73 (T)	28.3 (31.8) 35	28 (31.5) 34.5	0.99	0.43 (0.5) 0.57
M. pellita	0.13-0.7 (T)	0.06-0.13 (T)	0.03 (T)	0.53-0.6 (T)	0.93-1.6 (T)	0.6 (T)	28.5 (30.8) 32.5	29.3 (31.2) 33.3	0.98	0.53 (0.62) 0.67
S. acuminatus	0.23-0.4 (T)	0.13-0.37 (T)	0.18-0.32 (T)	0.13-0.23 (T)	0.83-1.11 (T)	0.32-0.46 (T)	28 (32.4) 36	23.6 (26.2) 29.7	1.23	0 (0.26) 0.30
S. albus	0.46-0.69 (T)	0.13-0.4 (T)	0.05-0.13 (T)	0.27-0.32 (T)	1.01-1.29 (T)	0.37-0.42 (T)	30 (35.9) 40	28 (31.6) 34	1.14	0 (0.21) 0.31
S. apoensis	0.4-0.6 (T)	0.3-0.5 (T)	0.2-0.4 (T)	0.2-0.4 (T)	1.06-1.33 (T)	0.4-0.6 (T)	25 (27.8) 30	23.7 (26.1) 29	1.14	0.23 (0.28) 0.38
Elmer 11795	–	–	–	–	–	–	36 (42.2) 50	33 (37.1) 44	1.1	0.31 (0.33) 0.42
S. auricomus	0.6 (S)	0.2-0.5 (S)	?	?	1.11-1.33 (S)	0.3-0.7 (S)	21 (24) 27	21 (22.6) 24	1.06	0.23 (0.27) 0.41
S. auritus	0.8-1.2 (S)	0.2 (S)	?	?	1.99 (S)	0.8 (S)	29 (29.5) 30	32 (33) 34	0.89	0.25 (0.33) 0.45
S. crassifolius	0.3 (S)	0.2-0.5 (S)	?	?	1.4-1.5 (S)	0.3-0.7 (S)	26.5 (29.2) 31.5	23 (26.0) 28.5	1.13	0 (0.21) 0.29
S. dubius	0.5-0.7 (S)	0.3-0.5 (S)	?	?	1.6 (S)	0.7-0.9 (S)	24.5 (27.4) 29	21.5 (25.4) 28	1.08	0 (0.31) 0.36
S. ferrugineus	0.2-0.4 (T)	0.26-0.6 (T)	0.06-0.13 (T)	0.3-0.4 (T)	1.06-1.2 (T)	0.46-0.53 (T)	30 (34.0) 37.3	25.7 (28.2) 31.8	1.2	0.17 (0.22) 0.26
S. gyrocarpus	?	?	?	?	?	?	23 (24.8) 27	21 (23) 25	1.08	0.4 (0.44) 0.49
S. harmandii	0.3-0.4 (S)	0.3-0.7 (S)	?	?	1.2-1.33 (S)	0.4 (S)	28 (30.5) 33	27.5 (29.1) 30.7	1.05	0 (0.18) 0.24
S. hirsutus	0.5-0.79 (T)	0.23-0.32 (T)	0.18-0.27 (T)	0.5-0.69 (T)	1.38-1.85 (T)	0.69-0.83 (T)	33.3 (34.8) 36.4	31.6 (33.3) 34.8	1.05	0.31 (0.37) 0.43
S. latibractea	0.55-0.69 (T)	0.32-0.6 (T)	0.09-0.42 (T)	0.28-0.6 (T)	1.7 (T)	0.69-0.79 (T)	29 (31.4) 34.3	29 (30.7) 33.6	1.02	0(0.17) 0.23
S. latistipulus	0.6-1 (S)	0.1-0.4 (S)	?	?	1.3-1.7? (S)	0.6-1.2 (S)	24 (27.5) 30.3	22 (25.9) 28	1.08	0.2 (0.3) 0.4
S. littoralis	0.23-0.46 (T)	0.13-0.37 (T)	0.27 (T)	0.18-0.4 (T)	1.02-1.38 (T)	0.45-0.69 (T)	26.4 (28.9) 30.9	26 (28.6) 30.6	1.01	0 (0.23) 0.29
S. macropterus	0.33-0.4 (T)	0.2-0.26 (T)	0.06-0.13 (T)	0.33-0.46 (T)	1-1.13 (T)	0.39-0.53 (T)	30.9 (32.8) 35.1	28.3 (30.2) 32.1	1.09	0.24 (0.36) 0.39
S. maingayi	0.26-0.4 (T)	0.2-0.7 (T)	0.2-0.3 (T)	0.2-0.4 (T)	0.93-1.46 (T)	0.4-0.6 (T)	28.4 (31.4) 33.9	27.5 (30.1) 32.1	1.04	0 (0.29) 0.36
S. merguensis	0.6 (S)	0.22-0.44 (S)	?	?	1.18 (S)	0.22-0.5 (S)	25 (27.2) 30	21 (25.5) 29	1.07	0 (0.32) 0.44
S. multiflorus	0.37-0.46 (T)	0.23-0.32 (T)	0.23-0.32 (T)	0.37-0.69 (T)	0.9-1.43 (T)	0.69-0.88 (T)	24.7 (28) 28.3	23.6 (25.4) 27	1.06	0.3 (0.33) 0.37
S. oblongifolius	0.32-0.4 (T)	0.093-0.23 (T)	0.093 (T)	0.37-0.5 (T)	1.02-1.11 (T)	0.5-0.6 (T)	31.6 (34.4) 37.6	30.6 (33.4) 36.7	1.03	0 (0.18) 0.25
S. parviflorus	0.44-0.85 (T)	0.27-0.5 (T)	0.06(-0.11) (T)	0.55-0.8 (T)	1.38-1.97 (T)	0.55-0.83 (T)	41.1 (43.5) 46.2	39.7 (42.5) 46.4	1.01	0.27 (0.3) 0.33
S. persicinus	0.26-0.6 (T)	0.13-0.3 (T)	0.2-0.4 (T)	0.26-0.4 (T)	1.13-1.66 (T)	0.46-0.8 (T)	33.6 (37.2) 40.8	31.6 (35.4) 39.2	1.05	0 (0.23) 0.3
S. pottingeri	0.4-0.5 (S)	0.5-0.7 (S)	?	?	1.44-1.55 (S)	0.4-0.6 (S)	30.9 (32.2) 33.4	26.7 (28.5) 29.7	1.03	0.25 (0.34) 0.41
S. pulcher	?	?	?	?	0.5 (S)	0.3 (S)	27.5 (31.7) 35	24.5 (28.4) 30.5	1.1	0.5 (0.58) 0.65
S. purpureus	0.093-0.13 (T)	0.13-0.23 (T)	0.093-0.13 (T)	0.27-0.37 (T)	0.69 (T)	0.37-0.42 (T)	22 (24) 26	23 (23.8) 25	1	0-0.12
S. ridleyi	?	?	?	?	?	?	26 (28) 30	26 (26.5) 27	1.06	0.38
S. sanguineus	0.18-0.23 (T)	0.23-0.4 (T)	0.18-0.27 (T)	0.23-0.37 (T)	1.01-1.06 (T)	0.46 (T)	28.8 (32.7) 32.6	27.3 (28.8) 30.5	1.16	0.43 (0.44) 0.47
S. suberectus	0.46-0.6 (T)	0.18-0.23 (T)	0.093-0.18 (T)	0.5-0.64 (T)	1.29-1.38 (T)	0.51-0.69 (T)	32.5 (40.3) 44.5	32.5 (36.3) 40	1.12	0.29 (0.33) 0.37
S. viridis	0.3-0.53 (T)	0.1-0.2 (T)	0.13-0.26 (T)	0.33-0.66 (T)	1.1-1.2 (T)	0.55-0.66 (T)	30.1 (31.6) 33.6	28.9 (29) 30.3	1.09	0 (0.23) 0.31

cies in Table 5.1 were calculated from the extremes of all pertinent samples: the mean of the minimums is the lower value, the mean of the maximums is the higher value; the middle value is the mean of the means. The terminology used in the descriptions follows Punt et al. (1994).

Specimens sampled

Butea monosperma (Lamarck) Taubert: Backer 9536 (Java, L), Brasong Watdahnahsahp 104 (Thailand, L), Dickason 6939bis (Burma, L), École des Jeunes Filles (Vietnam, P), Jacquemont 123 (India, P), Jacquemont 125 (India, P), Kostermans 24393 (Sri Lanka, L), Kostermans 28597 (Sri Lanka, L; TEM), Misra 1081 (India, L), Phengklai 3203 (Thailand, P), Poilane 11765 (Laos, P), Teijsmann 1860 (Hort. Bogor., L).

Butea superba Roxburgh: BKF 26128 (Thailand, L), BKF 30585 (Thailand, L), BKF 54411 (Thailand, L), Chevalier 31338 (Vietnam, P), Gamble s.n. (India, K), Harmand 308 (Laos, P), Harmand 334 (Laos, P), Hort. Bot. Calcutta s.n. (Hort. Bot. Calc., L), Maxwell 75-49 (Thailand, L), Mooney 945 (India, K).

Kunstleria forbesii Prain: Elmer 13105 (Philippines (Palawan), U), Elmer 21322 (Borneo, BO), Forbes 3241 (Borneo, L; TEM), King's collector 3094 (Malay Peninsula (Perak), K), Nieuwenhuis 1282 (Borneo, BO), SAN 89034 (Borneo, L).

Kunstleria geesinkii Ridder-Numan & Kornet: SAN 31574 (Borneo, K), SAN 52944 (Borneo, K), SAN 54866 (Borneo, K).

Kunstleria kingii Merrill: Kunstler 6870 (Malay Peninsula, K), Kunstler 6935 (Malay Peninsula, K).

Kunstleria philippinensis Merrill: Wenzel 818 (Philippines, US).

Kunstleria ridleyi Prain: Ahmed 1226 (Hort. Sing., BO), Austin Cuadra A 1043 (Borneo, BO), Elmer 21494 (Borneo, L), Derry 1006 (Malay Peninsula, K), FRI 8360 (Malay Peninsula, L), FRI 10518 (Malay Peninsula, L), Hort. Bog. XVII.F.13 (Hort. Bog., L), Hort. Bog. XVII.F.36 (Hort. Bog., L), Kostermans 6569 (Borneo, L), Maingay 1168 (K.D. 609) (Malay Peninsula (Malacca), K), Puasa BNB 1429 (Borneo, BO).

Kunstleria sarawakensis Ridder-Numan & Kornet: S 18036 (Borneo, K).

Meizotropis buteiformis Voigt: Koelz 28394 (India, L; TEM), Lobb 29 (India, BM), Ludlow c.s. 20928 (Bhutan, BM), Stainton 1787 (Nepal, BM), Stainton c.s. 5426 (Nepal, BM), Voigt s.n. (Assam, C).

Meizotropis pellita (Hooker f. ex Prain) Sanjappa: Lambert 1 (India, K), Polunin c.s. 1980 (Nepal, BM; TEM), Sprawson s.n. (India, K), Stainton c.s. 45 (Nepal, BM).

Spatholobus acuminatus Bentham: Forrest 13624 (Burma, K), Kerr 12317 (Thailand, K), Kerr 18291 (Thailand, K), Lace 2788 (Burma, K), Lister 293 (Bangladesh (Chittagong), K), Maxwell 75-137 (Thailand, L; TEM), Parkinson 393 (Andamans, K), Poilane 14573 (Cambodia, P), Put 1504 (Thailand, K), Wall 5907 (Burma, K).

Spatholobus albus Wiriadinata & Ridder-Numan: S 32363 (Borneo, L; TEM), S 41502 (Bor-neo, L).

Spatholobus apoensis Elmer: Beccari P.B. 2723 (Borneo, FI), Elmer 11795 (Philippines, L), Elmer 13300 (Philippines, L), Haviland 2897 (Borneo, L).

Spatholobus auricomus Ridder-Numan: S 22446 (Borneo, L).

Spatholobus auritus Ridder-Numan: Haniff 3703 (Borneo, K).

Spatholobus crassifolius Bentham in Miquel: Griffith K.D.1826 (India, K), Wallich 5913B (India, K).

Spatholobus dubius Prain: Curtis s.n., April 1893 (Malay Peninsula, K), Kunstler 7585 (Malay Peninsula (Perak), K).

Spatholobus ferrugineus (Zollinger and Moritzi) Bentham in Miquel: Backer 17437 (Java, L), van Balgooy 3977 (Celebes, L), FRI 4116 (Malay Peninsula (Malacca), L), van Niel 3868 (Borneo, L), van Niel 4397 (Borneo, L), PNH 12513 (Philippines, L), Richards 2542 (Borneo, L), S 16511 (Borneo, L), SAN 30919 (Borneo, L), Teijsmann s.n. (Sumatra, L).

Spatholobus ferrugineus var. *acutis* Ridder-Numan: Elsener H80 (Borneo, L), Franken & Roos 278 (Sumatra, L).

Spatholobus ferrugineus var. *sericophyllus* Ridley: Goodenough s.n. (Singapore, SING).

Spatholobus gyrocarpus Bentham in Miquel: KL 2758 (Malay Peninsula, L).

Spatholobus harmandii Gagnepain: Harmand s.n. (Laos, P), How 71016 (China (Hainan), P), Maxwell 71-733 (Thailand, L), Poilane 14064 (Cambodia, P).

Spatholobus hirsutus Wiriadinata & Ridder-Numan: Endert 5162 (Borneo, BO), Kostermans 13652 (Borneo, L), Mogea 4085 (Borneo, L; TEM), Nieuwenhuis 1079 (Borneo, L), SAN 37132 (Borneo, K), SAN 76943 (Borneo, L), SAN 77125 (Borneo, L).

Spatholobus latibractea Ridder-Numan: S 45351 (Borneo, L; TEM), SAN 97466 (Borneo, L).

Spatholobus latistipulus Merrill: Elmer 21439 (Borneo, L), SAN 65877 (Borneo, L), Wiriadinata 3397 (Borneo, L).

Spatholobus littoralis Hasskarl: Backer 11016 (Java, L), Backer 17276 (Java, L), Blume s.n. (Java, L), Blume s.n. (Java, L), Blume s.n. (Java, L), BS 28217, BS 30947 (both Philippines, L), de Voogd 526 (Sumatra, BO), Winckel 325 (Java, L; TEM), Winckel 488b (Java, L).

Spatholobus macropterus Miquel: FRI 10529 (Malay Peninsula (Malacca), L), Elmer 13063 (Philippines, U), Elmer 21390 (Borneo, L), Elmer 21398 (Borneo, L), Forbes 3195 (Sumatra, L), Forbes 3243 (Sumatra, L; TEM), Kostermans 6872 (Borneo, L), PNH 23077 (Philippines, L), Rutten 84 (Borneo, U), SAN 96941, SAN 97328, SAN 97452 (all Borneo, L).

Spatholobus maingayi Prain ex King: FB 21474 (Philippines, P), Kunstler 10428 (Malay Peninsula, L; TEM), FRI 8430 (Malay Peninsula, L), Maingay 611 (Malay Peninsula, L), PNH 2911 (Philippines, L), SAN 30594 (Borneo, L), SAN 50484 (Borneo, L), SAN 76900 (Borneo, L), SAN 76946 (Borneo, L), Santos 4064 (Philippines, L).

Spatholobus merguensis Prain: Parkinson 1690 (Burma, K).

Spatholobus multiflorus Wiriadinata & Ridder-Numan: Maidin 7331 (Borneo, SING), S 32370 (Borneo, L), SAN 96975 (Borneo, L), SAN 110486 (Borneo, L).

Spatholobus oblongifolius Merrill: Clemens 30236 (Borneo, L), Hose 486 (Borneo, L), Hose 632 (Borneo, L), SAN 43060 (Borneo, L), SAN 55708 (Borneo, L), SAN 65944 (Borneo, L), SAN 97669 (Borneo, L), SAN 99420 (Borneo, L), SAN 100088 (Borneo, L; TEM).

Spatholobus parviflorus (Roxburgh ex DC.) O. Kuntze: d'Alleizette s.n, June 1909 (Cambodia, L), d'Alleizette 1799 (India, L), BKF 37911 (Thailand, L), Garrett 1353 (Thailand, L; TEM), J.D. Hooker s.n. (India, L), Kerr 20485 (Thailand, K), Koelz 19167 (India, L), Kostermans 1171 (Thailand, L), Maxwell 74-746 (Thailand, L), Smitinand 4815 (Thailand, L).

Spatholobus persicinus Ridley: Endert 5410 (Borneo, BO), Haviland & Hose 3276K (Borneo, K), Kostermans 7108 (Borneo, L; TEM), Motley 734 (Borneo, K), Nooteboom 4374 (Borneo, L), SAN 26779 (Borneo, K).

Spatholobus pottingeri Prain: BKF 37401 (Thailand, L), Forrest 12251(China, L), Henry 11771C (China, K), Poilane 20341 (Laos, P), Spire 1067 (Laos, P).

Spatholobus pulcher Dunn: Henry 12780A bis, Henry 12780B, Henry 13331 (all China, K).

Spatholobus purpureus Bentham ex Prain: Stocks s.n. (India, K; TEM).

Spatholobus ridleyi Prain ex King: Forbes 3195 (Sumatra, L), Ridley 6401 (Singapore, K).

Spatholobus sanguineus Elmer: Cuming 945 (Philippines, L), Elmer 17560 (Philippines, L; TEM), SAN 32504 (Borneo, L), SAN 69888 (Borneo, L), SAN 88852 (Borneo, L), SAN 100240 (Borneo, L), Wenzel 989 (Philippines, L).

Spatholobus suberectus Dunn: Bor 5067 (Assam, K), Henry 11977A (China, L; TEM), Henry 13698 (China, BM), Spire 453 (Laos, P).

Spatholobus viridis Wiriadinata & Ridder-Numan: Austin Cuadra A 1038 (Borneo, K), Kostermans 6586 (Borneo, L), Kostermans 21654 (Borneo, L), Meijer 2458 (Borneo, L), SAN 31372 (Borneo, K), SAN 66270 (Borneo, L; TEM), SAN 97423 (Borneo, L), SAN 97594 (Borneo, L).

RESULTS

The pollen of the four genera differs from each other in ectoaperture morphology, exine ornamentation, exine stratification, and slightly in size. Infrageneric variation is small in *Butea* and *Meizotropis*, both consisting of only two species, and in *Kunstleria*, of which six species were available. *Spatholobus* pollen is more diverse: three types, one of which has three subtypes, could be recognised. Pollen descriptions of *Butea*, *Kunstleria*, *Meizotropis*, and of the infrageneric groups in *Spatholobus* are given below. Data of individual species are listed in Table 5.1. A summary of a number of the pollen characters of *Butea*, *Kunstleria* and *Meizotropis* and of the types and subtypes of *Spatholobus* is given in Table 5.2.

Table 5.2. Summary of pollen characters of *Butea*, *Kunstleria*, *Meizotropis*, and *Spatholobus*.

Pollen types	size (P × E)	A/E	colpus ends	ornamentation
Butea	38.3 × 35.3	0.4	obtuse, never fused	coarsely perforate to microreticulate
Kunstleria	23.8 × 21.2	0.45	acute, never fused	rugulate to fossulate
Meizotropis	31.1 × 31.4	0.56	obtuse, never fused	coarsely fossulate-microreticulate to verrucate
Spatholobus				
Type I	43.5 × 42.5	0.3	acute, never fused	psilate-perforate to fossulate, rarely rugulate
Type II	32.5 × 30.5	0.28	acute to obtuse, never fused	coarsely perforate-fossulate, micro-reticulate, often deviating verrucate
Type III				
Subtype IIIa	31.8 × 28.7	0.37	acute, never fused	psilate-perforate to fossulate
Subtype IIIb	30.2 × 27.9	0.26	acute, often fused	psilate-perforate to fossulate, coarsely perforate, sometimes rugulate or deviating
Subtype IIIc	32.1 × 30.4	0.36	acute, rarely obtuse, sometimes fused	psilate to psilate-perforate to fossulate, nearly microreticulate

(continued)

Pollen types	infratectal structure	foot layer / endexine
Butea	short columellae / granule-like columellae	1/4–1/8
Kunstleria	(irregular) columellae	1/2–1 1/2
Meizotropis	indistinct / granular / granule-like columellae	less than 1/10
Spatholobus		
Type I	coarsely granular / granule-like columellae	less than 1/8
Type II	irregular short columellae / (coarsely) granular	1/5–1 1/2
Subtype IIIa	well-spaced, sometimes irregular, columellae	up to 1/3 (or equal)
Subtype IIIb	(densely spaced) short (irregular) columellae / granule-like columellae / (coarsely) granular	1/4–2
Subtype IIIc	dense irregular columellae / granule-like columellae / (coarsely) granular	1/6–1 1/3

General description

Pollen grains (sub)isopolar, tricolporate, sometimes (para)syncolporate monads.

Ectoapertures usually distinct colpi; with obtuse or acute ends. Endoapertures lalongate to isodiametric, or sometimes lolongate pori; lateral margins obtuse; costae absent.Exine tectate. Nexine thinner to distinctly thicker than sexine. Foot layer from less than 1/10 to twice as thick as endexine. Infratectum usually consisting of indistinct, short, irregular granule-like columellae or (coarse) granules, sometimes more distinctly columellate.

Ornamentation mostly psilate-perforate, coarsely perforate, microreticulate, sometimes rugulate, fossulate or verrucate.

Butea (Plate I: a–d)

Pollen grains isopolar. P = (34.3–)38.3(–42) μm, E = (31.8–)35.3(–38.3) μm. P/E = 1.07–1.11. Equatorial outline obtusely triangular with convex sides; meridional outline subcircular to obtusely rhombic.

Ectoapertures 5–10 μm wide, A/E = (0.3–)0.4(–0.51); colpus ends obtuse; colpus membrane completely covered with granules or granules only along the mesocolpia. Endoapertures lalongate to isodiametric pori, 4–8 × 4–10 μm.

Exine 0.93–1.44 μm. Nexine about as thick as sexine, 0.4–0.8 μm. Endexine 0.26–0.53 μm, thickening towards colpi; foot layer, 0.06–0.13 μm, 1/4–1/8 of the endexine, discontinuous. Infratectum distinct, 0.13–0.44 μm, gradually thinning towards colpi; consisting of rather loosely arranged short columellae or granule-like columellae. Tectum 0.2–0.55 μm.

Ornamentation at the mesocolpia microreticulate (Plate I b), with one or two, rarely up to seven, free sexine elements visible in each lumen, to coarsely perforate, at the apocolpia perforate (Plate I a) to fossulate, along the colpi sometimes psilate.

Butea monosperma and *B. superba* have similar pollen. The most distinctive features, compared with that of *Kunstleria*, *Meizotropis* and *Spatholobus*, are the coarsely perforate to microreticulate ornamentation, the obtuse colpus ends that are never fused at the poles, an infratectal structure with loosely arranged columellae and a foot layer which is usually less than 1/8 of the endexine. In general, the data in the literature agree with the above description; only slight differences in grain size and ornamentation occur. Murthy (1992) mentioned a difference in grain size: P = 35–40 μm for *B. monosperma*, and P = 44–55 μm for *B. superba*. We found no consistent difference in size between the two species; one sample of *B. superba* contained relatively large grains (Chevalier 31338: P = 46–52 μm).

Kunstleria (Plate I: i–o)

Pollen grains isopolar. P = (20.2–)23.8(–30) μm, E = (17.5–)21.2(–24) μm. P/E = 1.07–1.25. Equatorial outline circular to obtusely triangular with slightly convex sides; meridional outline circular to obtusely rhombic or slightly elliptic.

Ectoapertures (1–)2–5 μm wide, A/E = (0.32–)0.45(–0.58); colpus ends acute; colpus membrane completely covered with granules. Endoapertures isodiametric to lalongate pori, 4–6(–8) × 4–9(–14) μm.

Exine 0.83–1.17 µm. Nexine thinner than sexine, 0.22–0.53 µm. Endexine 0.1–0.4 µm, thickening considerably towards colpi; foot layer 0.11–0.24 µm, about half as thick to twice as thick as the endexine, continuous, but with fine channels, thinning towards colpi. Infratectum distinct, 0.1–0.44 µm, gradually thinning towards the colpi, consisting of (irregular) columellae. Tectum (0.16–)0.2–0.66 µm.

Ornamentation at the mesocolpia finely rugulate to fossulate, at the apocolpia fossulate.

The pollen grains of the six species studied are more or less similar. Unfortunately, it was not possible to study the other two species, due to lack of flowering material. Characteristic features of this genus are a finely rugulate to fossulate ornamentation, acute colpus ends which are never fused, an infratectal layer of columellae, and a foot layer which is equal to or half as thick as the endexine.

The short description of *K. philippinensis* by Bulalacao (1990: "rugulate fossulate with distinct often parallel muri sometimes anastomosing") generally agrees with the above description, except for slight differences in grain size.

We noticed that in *K. forbesii* the larger anthers yielded abundant pollen with a well-defined ornamentation and clear exine structure, whereas the smaller anthers contained fewer and smaller grains, irregular in shape and with a poorly developed ornamentation. A more comprehensive comparison of pollen from small and large anthers would be interesting.

Meizotropis (Plate I: e–h)

Pollen grains (sub)isopolar. P = (28.3–)31.3(–35) µm, E = (28–)31.4(–34.5) µm. P/E = 0.98–0.99. Equatorial outline circular to obtusely triangular with slightly convex sides; meridional outline circular, with one apocolpium sometimes flattened.

Ectoapertures short, indistinct colpi, 3–5(–10) µm wide, A/E = (0.43–)0.56(–0.67); colpus ends obtuse; colpus membrane with granules along the mesocolpia. Endoapertures isodiametric to lalongate, sometimes lolongate pori, 5–7(–14) × 5–7 µm.

Exine 0.93–1.6 µm. Nexine distinctly thicker than sexine, 0.6–0.73 µm. Endexine 0.53–0.7 µm, not thickening towards colpi; foot layer 0.03–0.06 µm, less than 1/10 of the endexine, discontinuous. Infratectum indistinct, 0.06–0.13 µm, gradually thinning towards the colpi, consisting of short granule-like columellae or granules. Tectum 0.13–0.7 µm.

Ornamentation coarsely fossulate-microreticulate to verrucate, with relatively large verrucae along the colpi and at the apocolpia.

The pollen of *M. buteiformis* is fossulate to verrucate while that of *M. pellita* is mostly verrucate, but in other characters the pollen of both species is similar. The colpus ends are obtuse and never fused at the poles; the infratectal layer is thin and consists of indistinct granule-like columellae; the foot layer is less than 1/10 of the endexine.

Meizotropis pollen is described in detail by Murthy (1989, 1992). Apart from the thickness of the exine (2–2.5 µm according to Murthy, but 0.93–1.6 µm in this study) and the maximum size of the grains (up to 39 µm versus 35 µm), his results agree with our data.

Spatholobus

Spatholobus **Type I** (Plate II: a–c)

Pollen grains (sub)isopolar. P = (41.1–)43.5(–46.2) μm, E = (39.7–)42.5(–46.4) μm. P/E = 1.01. Equatorial outline circular to obtusely triangular with slightly convex sides; meridional outline circular, with one apocolpium sometimes flattened.

Ectoapertures distinct colpi, 3–10 μm wide, A/E = (0.27–)0.3(–0.33); colpus ends acute; colpus membrane completely covered with granules or granules only along the mesocolpia. Endoapertures lalongate pori, once observed as extending to the colpus ends, 6–7 × 12–13 μm.

Exine 1.38–1.97 μm. Nexine about as thick as sexine, 0.55–0.83 μm. Endexine 0.55–0.8 μm, slightly thickening towards colpi; foot layer 0.05(–0.11) μm, less than 1/8 of the endexine (Plate II c), continuous. Infratectum indistinct, 0.27–0.5 μm, slightly thinning towards the colpi, consisting of coarse granules, or irregular columellae with a granular structure. Tectum 0.44–0.85 μm.

Ornamentation psilate-perforate (Plate II b) to fossulate, rarely rugulate.

Species included: *S. parviflorus* (Plate II a–c). Pollen of *S. parviflorus* differs from that of most other *Spatholobus* species by its relatively large grain size (P × E = 41.1–46.2 × 39.7–46.4 μm) and the very large endexine/foot layer ratio (the foot layer is less than 1/8 of the endexine; Plate II c). Although *S. suberectus* and *S. persicinus* also have large pollen grains (32.5–44.5 × 28–45 μm and 33.6–40.8 × 31.6–39.2 μm respectively) and *S. suberectus* (Plate V g) also has a thin foot layer compared to the endexine (1/5 of the endexine), in all other aspects they are better placed in other groups: *S. suberectus* in Subtype IIIc, because of the thin and indistinct infratectal layer, and *S. persicinus* in Subtype IIIb on the grounds that the infratectal layer has short, granule-like columellae and the foot layer is thicker than in *S. parviflorus*. The ornamentation of *S. parviflorus* pollen is like that of various other *Spatholobus* species: psilate-perforate to fossulate, the colpus ends are acute and never fused, the infratectal layer consists of coarse granules or granule-like columellae.

In the literature, *S. parviflorus* is usually treated as a *Butea*. The pollen, however, is rather different from that of *Butea*, which is smaller, and has a more open tectum (coarsely perforate or microreticulate) and a columellate infratectal layer. The only similarity is the very thin foot layer compared to the endexine.

Murthy (1992) also gave a description of *S. parviflorus* (as *B. parviflora*). There are only minor differences: exine thickness 3 μm versus up to nearly 2 μm in the present study, and endoaperture width 18 μm versus up to 13 μm.

Spatholobus **Type II** (Plate II: d–m)

Pollen grains (sub)isopolar. P = (24–)32.5(–33.9) μm, E = (22–)30.5(–32.1) μm, P/E = 1.01–1.14; the type specimen of *S. apoensis* (Elmer 11795) possesses larger grains, P = (36–)42.2(–50) μm, E = (33–)37.1(–44) μm. Equatorial outline circular to obtusely triangular, with gaps at the apertures; meridional outline circular or elliptic to obtusely rhombic with slightly convex sides.

Ectoapertures usually distinct colpi, (1–)2–5 μm wide, A/E = (0.18–)0.29(–0.42); colpus ends obtuse to acute; colpus membrane completely covered with coarse granules (Plate II e) or granules only along the margins (Plate II a). Endoapertures lalongate pori, 3–9 × 2–10 μm.

Exine 0.93–1.7 μm. Nexine about as thick as sexine, 0.4–1.2 μm; endexine 0.18–0.6 μm, gradually thickening towards colpi; foot layer 1/5 to 1 1/2 of the endexine, 0.2–0.4 μm, continuous or discontinuous. Infratectum not very distinct, 0.1–0.7 μm, consisting of irregular columellae and/or coarse granules. Tectum 0.23–1.0 μm.

Ornamentation microreticulate with one free sexine element per lumen (Plate II h) or coarsely perforate-fossulate (Plate II k, l); more or less verrucate in deviating grains.

Species included: *S. apoensis* (Plate II d–f), *S. latistipulus* (Plate II k, l), *S. littoralis* (Plate II i), and *S. maingayi* (Plate II g, h, j, m). The pollen of these species is different from that of all other *Spatholobus* species because of their coarsely fossulate or perforate to microreticulate ornamentation, which is more like that in *Butea* and *Meizotropis*. The colpus ends are often obtuse, and the colpus membrane is often completely covered with granules. The foot layer and endexine are equally thick.

In this group, pollen grains with a deviating, verrucate to very coarsely fossulate ornamentation (Plate II e, f) were frequently observed among normal grains.

Spatholobus Type III

Pollen grains (sub)isopolar, tricolporate, sometimes (para)syncolporate. P = (21–)30.9(–44.5) μm, E = (21–)28.9(–40) μm. P/E = 0.89–1.23. Equatorial outline circular to obtusely triangular with slightly convex sides; meridional outline circular to elliptic to rhombic, slightly angular at the apertures.

Ectoapertures distinct colpi, often equatorially constricted, (1–)2–5(–10) μm wide, A/E = (0–)0.32(–0.58); colpus ends acute, sometimes obtuse, fused or not; colpus membrane with granules only along the mesocolpia or completely covered with granules. Endoapertures lalongate to circular pori, sometimes lolongate and extending to the colpus ends, 3–10 × 5–15 μm.

Exine (0.5–)1.0–1.99 μm. Nexine thinner than or as thick as sexine, 0.22–0.9 μm. Endexine 0.13–0.69 μm, slightly thickening towards colpi; foot layer thinner than endexine or slightly thicker, 0.05–0.6 μm, continuous or discontinuous. Infratectum distinct or not, consisting of well-spaced columellae, coarse granules or irregular columellae.

Ornamentation psilate, psilate-perforate to fossulate or microreticulate

Three subtypes based on relatively small differences in exine structure, ornamentation, and ectoaperture morphology (colpi fused or not) are described.

Spatholobus Subtype IIIa (Plate III: a–i, l–n)

Pollen grains tricolporate. A/E = (0.17–)0.37(–0.49); colpus ends never fused; colpus membrane usually with granules only along the mesocolpia, sometimes completely covered with granules. Endoapertures lalongate pori, sometimes lolongate and reaching the colpus ends (*S. macropterus* and *S. hirsutus*).

Exine 1.0–1.2(–1.85) µm. Nexine about as thick as sexine, 0.39–0.83 µm; endexine 0.23–0.69 µm; foot layer (0.06–)0.18–0.27 µm, up to 1/3 of the endexine or equal, continuous, with channels (Plate III d, n) or slightly discontinuous (*S. macropterus*; Plate III h). Infratectum distinct, 0.2–0.4 µm, consisting of distinct, well-spaced (Plate III c), sometimes irregular (Plate III d), columellae. Tectum 0.18–0.4 µm (0.79 µm in *S. hirsutus*).

Ornamentation psilate-perforate to fossulate, sometimes slightly microreticulate or slightly microrugulate (*S. hirsutus*; Plate III l).

Species included: *S. ferrugineus* (Plate III a–d), *S. gyrocarpus* (Plate III e), *S. hirsutus* (Plate III i, l–n), *S. macropterus* (Plate III f–h), and *S. sanguineus*. They are grouped together because of the presence of distinct columellae, the ornamentation varying from psilate-perforate to fossulate or even slightly microreticulate, and the colpi never being fused at the poles. The foot layer is up to 1/3 of the thickness of the endexine. Within this group *S. hirsutus* and *S. macropterus* deviate slightly because of the lolongate endoapertures; the tectum of *S. hirsutus* is thicker than that of other species and the ornamentation is microrugulate or microreticulate; *S. macropterus* sometimes has a thin foot layer (up to 1/5 of the endexine).

No differences were found between the pollen of the three varieties within *S. ferrugineus* recognised by Wiriadinata and Ridder-Numan (1985).

Spatholobus **Subtype IIIb** (Plate III: j, k; Plate IV; Plate V: a–c)

Pollen grains tricolporate or (para)syncolporate (Plate IV e; Plate V a). A/E = (0–)0.29(–0.41); colpus ends acute and often fused at the poles; colpus membrane usually with granules only along the mesocolpia, sometimes completely covered with granules. Endoapertures lalongate to circular, sometimes slightly lolongate pori, sometimes lolongate and reaching the colpus ends (*S. acuminatus*, *S. persicinus*, *S. viridis*).

Exine (0.69–)1.33–1.66 µm. Nexine thinner than or as thick as sexine, 0.23–0.7 (–0.9) µm; endexine 0.13–0.69 µm; foot layer 0.05–0.4 µm, 1/4 of the endexine to twice as thick, continuous or discontinuous (Plate IV o). Infratectum not very distinct (Plate IV h), (0.09–)0.5(–0.7) µm, consisting of (coarse) granules (Plate IV g), granule-like columellae (Plate IV k, l) or short (irregular) columellae(Plate IV o). Tectum (0.093–)0.3–0.7 µm.

Ornamentation psilate (Plate IV f, n), psilate-perforate (Plate III j) to fossulate (Plate V c) or coarsely perforate, sometimes rugulate or deviating.

Species included: *S. acuminatus* (Plate IV a–d), *S. albus* (Plate IV e, f, h), *S. auricomus*, *S. crassifolius*, *S. dubius*, *S. harmandii* (Plate IV g), *S. multiflorus* (Plate III k, l), *S. oblongifolius* (Plate IV i–l), *S. persicinus* (Plate V a–c), *S. purpureus* (Plate IV m), *S. ridleyi*, and *S. viridis* (Plate IV n, o). The tectum in this group is less densely perforate than in Subtype IIIa, the colpi are often fused at the poles, the infratectal layer consists of rather indistinct columellae or granules (in *S. multiflorus* sometimes distinct columellae), and the foot layer is 1/2 to 2/3 of the endexine.

Of these species only *S. harmandii* is treated in the literature (Graham & Spencer Tomb, 1977; as *B. compur*, which should be *S. compar*, a synonym of *S. harmandii*). The description does not differ from that given above.

Spatholobus **Subtype IIIc** (Plate V: d–m)

Pollen grains tricolporate or sometimes (para)syncolporate. A/E = (0–)0.36(–0.65); colpus ends acute or sometimes obtuse (*S. auritus*: Plate V d), sometimes fused; colpus membrane with granules only along the mesocolpia or completely covered with granules. Endoapertures lalongate to circular pori.

Exine 1.18–1.99 µm (0.5 µm in *S. pulcher*, LM). Nexine thinner than or as thick as sexine, 0.22–0.8 µm. Endexine 0.55–0.69 µm; foot layer 0.09–0.6 µm, 1/6 to 1 1/3 of the endexine, continuous (in some species only the total nexine thickness is known). Infratectum indistinct, 0.18–0.7 µm, very dense (Plate V f, g), coarsely granular, or with irregular or granule-like columellae. Tectum 0.4–1.2 µm.

Ornamentation mostly psilate (Plate V i–l) to psilate-perforate, also fossulate, sometimes nearly microreticulate (Plate V h).

Species included: *S. auritus* (Plate V d), *S. latibractea* (Plate V h–j), *S. merguensis* (Plate V m), *S. pottingeri* (Plate V k), *S. pulcher* (Plate V l), and *S. suberectus* (Plate V f, g). The most striking features of Subtype IIIc are the relatively thick, sparsely perforate tectum (Plate V g, j), the comparatively thick endexine with a foot layer being 1/5 to equal to the endexine, and the relatively thin, dense and coarsely granular infratectal layer. Fused colpus ends occur sometimes in this subtype.

The first two species are slightly different because of their more open, microreticulate, ornamentation and the more obtuse colpus ends.

DISCUSSION

The basic pollen type in the Papilionoideae is spheroidal, tricolporate, finely reticulate, and with a well-defined endexine that is about as thick as the foot layer. The infratectal layer is columellate with well-spaced columellae, and about as thick as the nexine. The tectum is distinct and less than half as thick as the infratectum (Ferguson, 1984; Guinet & Ferguson, 1989). The variation found in most species of *Kunstleria* and *Spatholobus* Subtype IIIa fits the basic pollen type of the Papilionoideae. Pollen of *Butea* and *Meizotropis*, and of the other species of *Spatholobus* differ in some features from this basic type. Below, the pollen morphology of the genera studied is evaluated in relation to the trends in the Papilionoideae as defined by Ferguson and Skvarla (1981), Skvarla (1983), and Ferguson (1984), and compared to that of the Millettieae (Hazelhorst, 1981).

A phylogenetic analysis of the genus *Spatholobus*, and its allies *Butea* and *Meizotropis*, has been performed with PAUP (heuristic search, TBR options) (Swofford, 1991), using *Kunstleria* as outgroup. A datamatrix with 97 characters, 7 of which are pollen characters (Table 5.3), was used. The cladogram obtained from this analysis shows the distribution of these pollen characters (Fig. 5.1), which are evaluated below. See this thesis, Chapter 4, for a more detailed description of the methods for the phylogenetic analysis and the results and evaluation of the other characters.

Table 5.3. Datamatrix and coding of the pollen morphological characters for the phylogenetic analysis.

Species	1 ornamen- tation	2 colpi fused	3 footlayer / endexine	4 colpus ends	5 mesocolpial pouches	6 infra- tectum	7 A/E	(sub)type
B. monosperma	(2 4)	1	3	2	2	(1 2)	3	*Butea*
B. superba	(2 4)	1	?	2	2	(1 2)	3	*Butea*
K. forbesii	(1 4)	1	1	1	1	2	3	*Kunstleria*
K. geesinkii	(1 4)	1	?	1	1	2	3	*Kunstleria*
K. kingii	1	1	?	1	?	?	3	*Kunstleria*
K. philippinensis	(1 4)	1	?	1	?	?	4	*Kunstleria*
K. ridleyi	1	1	1	1	1	1	3	*Kunstleria*
K. sarawakensis	1	1	?	1	1	1	4	*Kunstleria*
M. buteiformis	(2 4 5)	1	2	2	2	(2 3)	4	*Meizotropis*
M. pellita	(2 5)	1	2	2	2	(2 3)	4	*Meizotropis*
S. acuminatus	(3 4)	2	1	1	1	2	1	IIIb
S. albus	3	2	3	1	1	(2 3)	2	IIIb
S. apoensis	2	1	1	(1 2)	1	(2 3)	1	II
S. auricomus	(3 4)	1	?	1	?	(2 3)	1	IIIb
S. auritus	3	1	?	2	?	4	3	IIIc
S. crassifolius	3	2	?	1	?	3	2	IIIb
S. dubius	3	2	?	1	?	3	1	IIIb
S. ferrugineus	(3 4)	1	3	1	1	1	2	IIIa
S. gyrocarpus	3	1	?	1	1	?	4	IIIa
S. harmandii	(3 4)	3	?	1	1	(2 3)	2	IIIb
S. hirsutus	(3 4)	1	(1 3)	1	1	(1 2)	3	IIIa
S. latibractea	(2 3)	2	1	1	1	(2 4)	2	IIIc
S. latistipulus	4	1	?	(1 2)	?	3	1	II
S. littoralis	(2 4)	2	(1 3)	(1 2)	1	(2 3)	2	II
S. macropterus	(3 4)	1	3	1	1	(1 2)	(1 3)	IIIa
S. maingayi	(2 4 5)	2	1	(1 2)	1	3	1	II
S. merguensis	3	2	?	1	1	(3 4)	1	IIIc
S. multiflorus	3	1	1	1	1	(1 2 4)	3	IIIb
S. oblongifolius	(3 4)	2	3	1	1	(2 3)	2	IIIb
S. parviflorus	(3 4)	1	2	1	2	(2 3)	1	I
S. persicinus	(2 3 4)	3	(1 3)	1	1	(2 3)	2	IIIb
S. pottingeri	3	1	?	1	1	(2 3 4)	3	IIIc
S. pulcher	3	1	?	1	2	?	4	IIIc
S. purpureus	3	2	3	1	1	(2 3)	2	IIIb
S. ridleyi	4	1	?	1	1	?	3	IIIb
S. sanguineus	3	1	(1 3)	1	1	(1 2)	4	IIIa
S. suberectus	(3 4)	1	3	1	1	(2 3 4)	3	IIIc
S. viridis	(3 4)	3	3	1	1	(2 3)	2	IIIb

1. Ornamentation:
    1 = rugulate
    2 = microreticulate / coarsely perforate
    3 = psilate-perforate
    4 = fossulate
    5 = verrucate
2. Colpi fused:
    1 = never
    2 = sometimes
    3 = often
3. Foot layer / endexine:
    1 = foot layer ≥ 1/2 of the endexine
    2 = foot layer < 1/8 of the endexine
    3 = foot layer > 1/8, but < 1/2 of the
        endexine

4. Colpus ends:
    1 = acute
    2 = obtuse
5. Mesocolpial pouches:
    1 = mostly present
    2 = mostly absent
6. Infratectal layer:
    1 = well-spaced columellae
    2 = irregular columellae
    3 = granules
    4 = indistinct / dense
7. A/E index:
    1 = 0.25 < mean < 0.35
    2 = mean < 0.25; max. = 0.35
    3 = mean ≥ 0.33; min. = 0.25
    4 = min. ≥ 0.4

The (sub)type is listed for each species, but not used for the phylogenetic analysis. ? = unknown, ( ) = polymorphic.

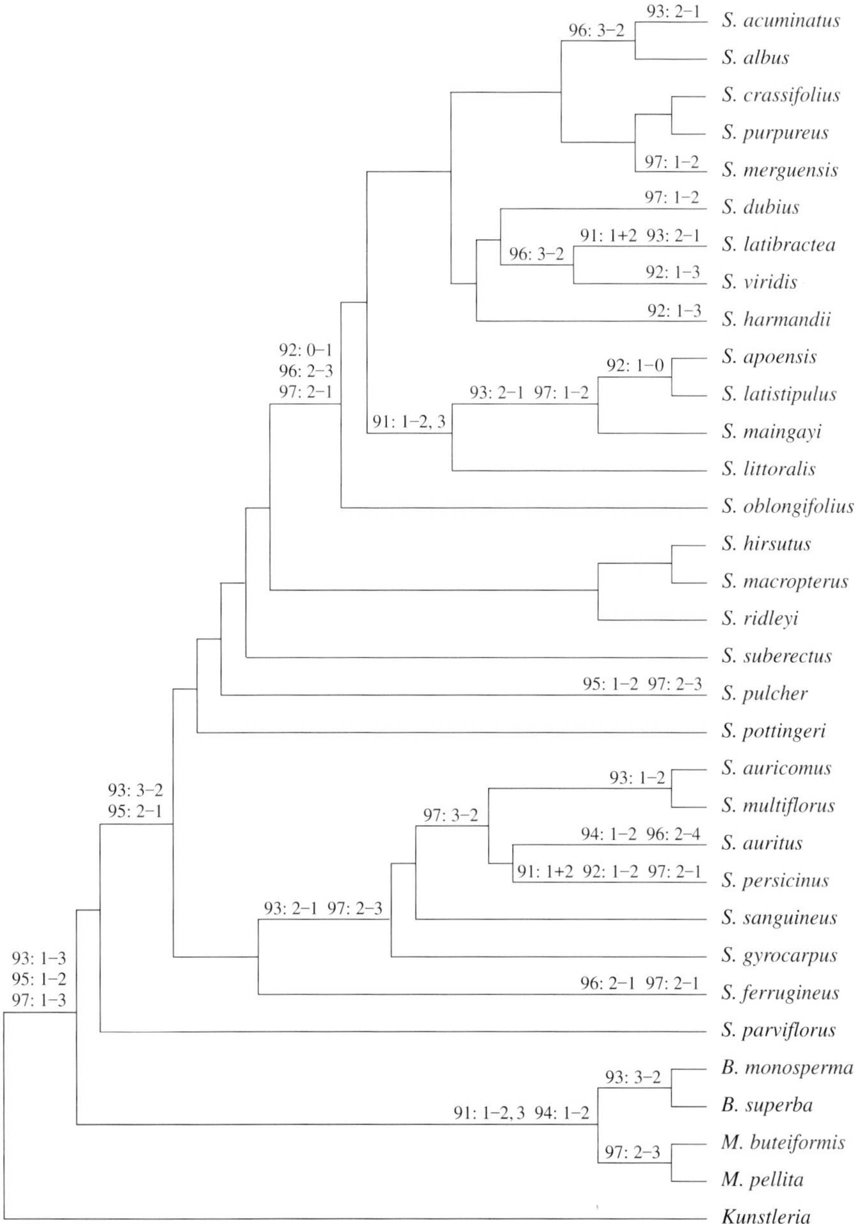

Figure 5.1. Cladogram of *Butea, Kunstleria, Meizotropis,* and *Spatholobus* showing character state changes of the pollen morphological characters. For character numbers and coding see Table 5.3. For reasons of comparison character numbers are identical to those in the original and complete datamatrix used for the phylogenetic analysis (in Chapter 2).

Evaluation of the pollen characters

Size and shape

The smallest pollen grains occur in *Kunstleria*, with P ranging from 18 to 30 µm. *Meizotropis* and *Butea* have grains of medium size: P = 28.3–35 µm and P = 34.3–42 µm respectively. Grains in *Spatholobus* vary in size from small to medium: P = 21–46.2 µm. Those of *S. parviflorus* always exceed 40 µm (P = 41.1–46.2 µm), and the next largest is *S. suberectus* (P = 32.5–44.5 µm). The sizes found in the four genera conform to those found in Papilionoideae pollen, which are averagely 30 µm, and usually not over 50 µm (Ferguson & Skvarla, 1981; Ferguson, 1984). Hazelhorst (1981) mentions averages of 20 to 40 µm in the tribe Millettieae.

Grain shape is not very variable. In equatorial view the grains are usually (sub)-spheroidal; in polar view they have a circular to obtusely triangular equatorial outline.

Because of the many overlapping values in grain size and the rather uniform grain shape, size and shape were not used in the phylogenetic analysis.

Apertures

All collections examined have mostly tricolporate pollen, in which the colpi are distinct, with acute or obtuse ends. In *Meizotropis*, the colpi are indistinct and short (Plate I a), as is often observed in the subtribe Erythrininae of the tribe Phaseoleae (Graham & Spencer Tomb, 1977). Within the tribe Phaseoleae, the reduction of colpi is considered to be derived (Ferguson, 1984; Ferguson & Skvarla, 1981), and the elongation and fusion of the colpi at the poles is generally also regarded as derived (Ferguson & Skvarla, 1981). In some species of *Spatholobus*, (para)syncolpy is frequently observed (*S. harmandii, S. persicinus*:Plate V a, *S. viridis*). In others, a small percentage of (para)syncolporate grains occurs in a predominantly colporate sample (Type II, Subtypes IIIb and c: Plate IV e), which is also known in the tribe Millettieae, e.g., in *Millettia pachyloba* and *Ostryoderris leucobotrya* (Hazelhorst, 1981).

In the cladogram (Fig. 5.1), the presence of fused colpi is restricted to the species in the upper half of the *Spatholobus* clade (from *S. oblongifolius* upwards). In the other species of *Spatholobus* (except *S. persicinus*), in the genera *Butea* and *Meizotropis*, and in the outgroup *Kunstleria* (para)syncolpy does not occur. The A/E index, which is the relative apocolpium size and indirectly a measure for the length of the colpi, was, as expected, large or medium-large in the species without (para)syncolpy. Especially in *Meizotropis*, large A/E values (> 0.5) occur due to the very short colpi. In the species with a (frequent) occurrence of (para)syncolporate pollen (in the cladogram from *S. oblongifolius* upwards), the A/E index is small or medium (mean A/E < 0.35). An exception is *S. ferrugineus* with a small relative apocolpium size and no (para)syncolporate pollen.

The presence of obtuse or acute colpus ends is discriminative between the genera. In *Butea* and *Meizotropis*, the colpus ends are obtuse (see also Murthy, 1989), while in the outgroup *Kunstleria* and almost all species of *Spatholobus* acute colpus ends are present. Obtuse colpus ends in *Spatholobus* occur only in *S. auritus*, placed in the first clade of *Spatholobus* in the cladogram, and (not unambiguously) the species with pollen of Type II.

The endoaperture is an isodiametric to lalongate pore; in *Meizotropis* the pore is sometimes lolongate. Incidentally, the pore extends to the colpus ends (observed several times in *S. acuminatus*, *S. hirsutus*, *S. macropterus*, *S. persicinus*, and once in *S. parviflorus* and *S. viridis*). As yet, no explanation for the origin of such meridionally elongated endoapertures has been found.

The colpus membrane is either completely covered with granules or smooth with granules along the mesocolpia. Unlike in the Millettieae (Hazelhorst, 1981), the presence of granules on the colpus membrane does not seem to be a taxonomically significant feature in the four genera studied here. Grains with a coarse ornamentation, as in *Spatholobus* Type II and *Meizotropis*, often have a colpus membrane completely covered with granules.

A remarkable feature in some species is the occurrence of protruding mesocolpial margins in the equatorial zone, in the literature referred to as 'mesocolpial pouches' (Ferguson & Skvarla, 1981) or 'equatorial constrictions of the colpi' (Bulalacao, 1990; Hazelhorst, 1981). According to Ferguson and Skvarla (1981), the presence of 'mesocolpial pouches' has no taxonomic significance at the tribal level in the Papilionoideae. At specific level, however, they may be useful (Ferguson, pers. comm.). In *Kunstleria* and most species of *Spatholobus*, mesocolpial pouches were present (Plate IV f, j); in most specimens of *Butea*, *Meizotropis*, *S. parviflorus* and *S. pulcher*, they were absent.

Exine stratification

Infrageneric variation is limited in *Butea*, *Kunstleria* and *Meizotropis*, but there is more variation in the much larger genus *Spatholobus*.

In *Kunstleria*, the nexine is not more than half of the sexine. In *Butea*, the nexine equals the sexine, while in *Meizotropis*, the nexine is distinctly thicker than the sexine. *Spatholobus* has a nexine about as thick as the sexine or slightly less.

In *Kunstleria*, the foot layer is about half the thickness of the endexine or more. In *Butea*, the foot layer is 1/8–1/4 of the endexine. Grains of *Meizotropis* have a very thin foot layer compared to the massive endexine (< 1/10). In *Spatholobus* the foot layer is well-developed in most species, and may be even thicker than the endexine. However, in *S. parviflorus*, the foot layer is about 1/10 of the endexine. In other species, the foot layer may be locally thin. The combination of a thin foot layer and a thick endexine is often found in the tribe Phaseoleae, e.g., *Erythrina flabelliformis*, *Mucuna stans*, *Pueraria phaseolides*, *Dumasia villosa*; in *Afgekia* and *Dahlstedtia* the foot layer is even absent (Ferguson & Skvarla, 1981). In the tribe Millettieae, this combination is present, but less frequent, e.g., in *Whitfordiodendron atropurpureum* (Ferguson & Skvarla, 1981). According to Ferguson and Skvarla (1981, 1983) and Ferguson (1984), a relatively thick endexine, associated with a reduced or even absent foot layer, is advanced in Papilionoideae pollen, especially in the Phaseoleae. The pollen of *Meizotropis* and *S. parviflorus* may therefore be considered as advanced in this respect. In the cladogram, the very thin foot layer is an autapomorphy for *Meizotropis* as well as *S. parviflorus* (but can equivocally be optimised as being the state at the base of the cladogram). The outgroup *Kunstleria* has a relatively thick foot layer, which represents the plesiomorphic state in the present study. In *Butea* and most species of *Spatholobus*, the foot layer is thinner, 1/8 to 1/2 of the endexine, but well-developed. In some groups, e. g., the species with pollen of *Spatholobus* Type II, the foot layer is as thick as in *Kunstleria*.

Ferguson and Skvarla (1981) also pointed out that in the Phaseoleae thickening of the endexine, thinning of the foot layer, and reduction of the colpi are associated with an elaboration of the exine sculpture. These features are all present in *Meizotropis*.

Towards the colpi, the tectum and infratectal layer become thinner, while the endexine becomes thicker as is usual in most of the dicotyledons. This thickening of the endexine is very distinct in grains of *Kunstleria* (Plate I o). In *Meizotropis*, however, no thickening at all was observed (Plate I g). The endexine of Millettieae pollen is thickened towards the apertures (Hazelhorst, 1981), which is usual in dicotyledons. In Phaseoleae pollen, however, it is not thickened when porate apertures are present (Ferguson, pers. comm.). This is consistent with the observations in the genus *Meizotropis*, which has very short colpi (nearly porate sometimes) and no extra thickening of the relatively thick endexine towards the aperture.

Infratectum structure

In all groups studied here, the thickness of the infratectal layer is from half to as thick as the tectum, except locally in case of conspicuous verrucae being present on the tectum. In *Kunstleria*, the infratectal layer consists of columellae, which are sometimes irregular. Such irregular columellae are also present in *Butea*. Pollen of *Meizotropis* has a very thin infratectum of short granule-like columellae or granules. The infratectal layer in *Spatholobus* consists of either well-spaced columellae (Subtype IIIa), densely packed irregular columellae or coarse granules. In most Millettieae studied by Hazelhorst, the infratectal layer is also dense and packed with irregular columellae; in *Leptoderris*, the columellae are distinct and well-spaced. In the tribe Phaseoleae, the infratectal layer is granular, or consists of distinct columellae, irregular columellae, or a combination of columellae and coarse granules (Ferguson & Skvarla, 1981; 1983). According to Ferguson (pers. comm.) the occurrence of granules may be associated with specialisation, which suggests that *Spatholobus* (usually with a granular infratectum except for Subtype IIIa) is more specialised than *Kunstleria* (usually with a columellate infratectum). When considering the presence of granules in the infratectal layer as advanced over a columellate structure, the species of *Spatholobus* Subtype IIIa have a plesiomorphic infratectum structure. Also, the presence of a well-developed foot layer in Subtype IIIa may be regarded as plesiomorphic, as is outlined before. In the cladogram the plesiomorphic nature of the columellate infratectum is confirmed, because in the outgroup *Kunstleria* a columellate infratectal structure is present. The columellate structure in *Spatholobus* (Subtype IIIa) occurs in the clade of *S. ferrugineus,* but Subtype IIIa is not restricted to this clade. In addition to species in the *S. ferrugineus* clade, *S. macropterus* and *S. hirsutus* have a columellate infratectum as well. In *Spatholobus*, the infratectal layer shows more often only granules in the second half of the cladogram (from *S. oblongifolius* upwards) indicating the advanced character of the granular infratectal layer.

Ornamentation

The ornamentation within *Kunstleria* is more or less microrugulate, sometimes slightly fossulate. A more open tectum is present in *Butea* and *Meizotropis*: *Butea* shows a microreticulate (lumina with free sexine elements) to coarsely perforate ornamentation, *Meizotropis* a coarsely fossulate-microreticulate to verrucate ornamentation (Plate I f). Most species of *Spatholobus* have pollen with a psilate-perforate, rarely fossulate ornamentation. Species of Type II show, as *Butea* and *Meizotropis*, a more open tectum, from perforate-fossulate and microreticulate (with usually one free sexine element per lumen) to deviatingly verrucate. This more open type of tectum is a synapomorphy for the genera *Butea* and *Meizotropis*, as well as for the clade consisting of *S. apoensis*, *S. latistipulus*, *S. littoralis*, *S. maingayi* (all Type II), and *S. persicinus* (Fig. 5.1). This would agree with the view of Ferguson and Skvarla (1981), who suggest that the presence of free standing elements in the lumina is indicative of some degree of specialisation. In Hazelhorst's survey, perforate ornamentation is the most common type in the tribe Millettieae; the reticulate type also occurs in the Millettieae, but is less common (e.g. *Afgekia*, *Callerya*, *Fordia unifoliolata*, *Leptoderris*, *Platycyamus*, *Platysepalum*, *Sarcodum scandens*, *Ostryocarpus*). In the Phaseoleae, the ornamentation is diverse, ranging from unsculptured (psilate-imperforate) to coarsely reticulate. Usually the ornamentation is more open than in the Millettieae (Ferguson & Skvarla, 1981; Ferguson, 1990; Graham & Spencer Tomb, 1977).

Verrucate pollen, as found in both *Meizotropis* and *Spatholobus* Type II, occurs in the Millettieae, e.g. in *Millettia extensa*, *Millettia teuszii* and *Platysepalum hirsutum* (Hazelhorst, 1981), as well as in the Phaseoleae, e.g., in *Mucuna* (Graham & Spencer Tomb, 1977; Ferguson, 1990; Horvat et al., 1992). Most likely such a verrucate ornamentation is an adaptation to certain types of pollinators (Ferguson, 1984, 1990; Guinet & Ferguson, 1989), and most probably developed independently in different groups. More or less similar ornamentation types are known from other legumes, e.g., Klitgaard and Ferguson (1992) showed a tentative correlation between pollination type (bats and/or moths) and verrucate ornamentation in *Browneopsis* (Caesalpinioideae).

Concluding remarks

In general, the pollen morphological characters in *Butea*, *Kunstleria* and *Meizotropis* proved to be rather uniform within the genera. Among the four genera and within *Spatholobus* differences were found in sculpturing of the tectum, ectoaperture morphology (colpi long or short, fused or not, colpus ends obtuse or acute), infratectal structure, and relative thickness of foot layer and endexine, which made it possible to use these pollen morphological characters for a phylogenetic analysis (see Table 5.3 and Fig. 5.1).

In the analysis, which is based on macromorphological, leaf anatomical and pollen morphological data, the four genera studied represent monophyletic groups. *Spatholobus* Type I pollen is found only in *S. parviflorus*. Type II characterises a single clade. However, the three subtypes of Type III proved to characterise unnatural groups, being mixed up in most clades.

The trends indicated by Ferguson and Skvarla (1981) and Ferguson (1984) are confirmed on an infrageneric scale in the results of the phylogenetic analysis so far: microreticulate was found to be apomorphic over psilate-perforate to fossulate or rugulate, but developed parallel in *Butea*/*Meizotropis* and *Spatholobus* Type II (clade with *S. apoensis*); short colpi as well as the fused colpi appeared more derived than long free colpi. The presence of obtuse colpus ends is a synapomorphy in *Butea* and *Meizotropis*, and occurs also in the clade of *Spatholobus* with pollen Type II. A granular infratectum was found in the more derived species of *Spatholobus*. A very thin foot layer combined with a relatively thick endexine is confined to *Meizotropis* and *S. parviflorus*, which are also derived in other characters.

Due to their inconstancy, the presence or absence of mesocolpial pouches was difficult to code. When used as 'often present' versus 'often absent' two species of *Spatholobus* (*S. parviflorus* and *S. pulcher*) and the species of *Butea* and *Meizotropis* all had the same state ('often absent'), while the outgroup and the rest of *Spatholobus* shared the state 'often present'.

The pollen of *Kunstleria* fits smoothly in the Millettieae, while that of *Meizotropis* and *Butea* can be easily accommodated in the Phaseoleae, a tribe generally regarded as more advanced. Pollen morphological characters, however, do not give a decisive answer for the place of *Spatholobus* in either the Millettieae or the Phaseoleae. In general, the presence of granule-like columellae in the infratectum, fused colpi, and a more open tectum (especially in *Spatholobus* Type II) indicate a higher level of specialisation, which may lead to the conclusion that a place in the Phaseoleae would be appropriate. *Spatholobus parviflorus,* in leaf and pod very similar to *Butea*, has pollen with a very thin foot layer and a very thick endexine as in *Butea*. However, the ornamentation is similar to the psilate-perforate ornamentation of most other species of *Spatholobus*. Pollen of *Spatholobus* Type II has a different kind of ornamentation, which is closer to that in the Phaseoleae, including that of *Butea* and *Meizotropis*. However, the ornamentation of *Spatholobus* Type II and the psilate-perforate or fossulate ornamentation found in most other *Spatholobus* species occur in the tribe Millettieae as well. In addition, the plesiomorphic nature of a well-developed foot layer do not support an advanced position. This suggests a position of *Spatholobus* basally in the Phaseoleae or in the transition zone between the Phaseoleae and the Millettieae as indicated by macromorphological characters. Possibly, further study of the ill-defined tribe of the Millettieae will give a more definitive answer.

ACKNOWLEDGEMENTS

First of all we would like to thank Mrs. Marieke Verkley-Hardenberg, Mr. Wim Star, and Mrs. Bertie Joan van Heuven, for their help in making the preparations and photographs. Further we would like to thank Prof. Dr. P. Baas, Dr. I.K. Ferguson, Prof. Dr. C. Kalkman, Dr. M.C. Roos, and an anonymous reviewer for reading and commenting on the manuscript. Thanks are also due to the directors of the herbaria of K, L, and P, who kindly gave permission to sample pollen from herbarium specimens.

This project is financially supported by the Dutch Science Organisation (Life Sciences, project nr. 805-440.041).

REFERENCES

Banik, S., Bhattacharya, K. and Chanda, S. 1986. Pollen flora of Dum Dum and adjacent areas of West Bengal with reference to the identification of dispersed palynomorphs. Trans. Bose Res. Inst. 49: 1–21.

Baudet, J.C. 1978. Prodrome d'une classification générique des Papilionaceae–Phaseoleae. Bull. Jard. Bot. Nat. Belg. 48: 183–220.

Bruneau, A., Doyle, J.J. and Doyle, J.L. 1995. Phylogenetic relationships in Phaseoleae: evidence from chloroplast DNA restriction site characters. In: M. Crisp and J.J. Doyle (eds.), Advances in Legume Systematics, part 7: Phylogeny: 309–330. Royal Botanic Gardens, Kew.

Bulalacao, L.J. 1990. Pollen morphology of the Philippine Leguminosae. National Museum Papers 1: 66–95.

Ferguson, I.K. 1984. Pollen morphology and systematics of the subfamily Papilionoideae (Leguminosae). In: W.F. Grant (ed.), Plant Biosystematics. Academic Press, Toronto. pp. 377–394.

Ferguson, I.K. 1990. The significance of some pollen morphological characters of the tribe Amorpheae and of the genus *Mucuna* (tribe Phaseoleae) in the biology and systematics of subfamily Papilionoideae (Leguminosae). Rev. Palaeobot. Palynol. 32: 317–367.

Ferguson, I.K. and Skvarla, J.J. 1981. The pollen morphology of the subfamily Papilionoideae (Leguminosae). In: R.M. Polhill and P.H. Raven (eds.), Advances in Legume Systematics, part 2: 859–896. Royal Botanic Garden, Kew.

Ferguson, I.K. and Skvarla, J.J. 1983. The granular interstitium in the pollen of subfamily Papilionoideae (Leguminosae). Amer. J. Bot. 70: 1401–1408.

Geesink, R. 1981. Tephrosieae. In: R.M. Polhill & P.H. Raven (eds.), Advances in Legume Systematics, part 1: 245–260. Royal Botanic Garden, Kew.

Geesink, R. 1984. Scala Millettiearum. Leiden Bot. Series 8. Leiden. 131 pp.

Graham, A. and Spencer Tomb, A. 1977. Palynology of *Erythrina* (Leguminosae: Papilionoideae): the subgenera, sections and generic relationships. Lloydia 40: 413–435.

Guinet, Ph. and Ferguson, I.K. 1989. Structure, evolution, and biology of pollen in Leguminosae. In: C.H. Stirton and J.L. Zarucchi (eds.), Advances in Legume Biology. Monogr. Syst. Bot. Missouri Bot. Gard. 29: 77–103.

Gupta, H.P. and Sharma, C. 1986. Pollen flora of North-West Himalaya. Lucknow. 181 pp.

Hazelhorst, H.-E. 1981. Pollen of the Millettieae. Unpublished manuscript, Leiden. 98 pp.

Horvat, F., Ferguson, I.K. and Stainier, F. 1992. Relation entre l'ultrastructure pollinique et la taxonomie de quatre espèces du genre *Mucuna* (Leguminosae–Papilionoideae). Rev. Palaeobot. Palynol. 72: 285–298.

Klitgaard, B.B. and Ferguson, I.K. 1992. Pollen morphology of *Browneopsis* (Leguminosae: Caesalpinioideae), and its evolutionary significance. Grana 31: 285–290.

Lackey, J.A. 1981. Phaseoleae DC. (1825). In: R.M. Polhill and P.H. Raven (eds.), Advances in Legume Systematics, part 1: 301–327. Royal Botanic Garden, Kew.

Mandal, S. and Chanda, S. 1981. Pollen atlas of Kalyani, West Bengal, with reference to aeropalynology. Trans. Bose Res. Inst. 44: 35–62.

Mondal, M.S. and Mondal, M. 1983. Pollen morphological studies of some indigenous floral drugs of India (Leguminosae). J. Econ. Tax. Bot. 4: 517–524.

Murthy, G.V.S. 1989. Pollen morphology of *Meizotropis* Voigt (Fabaceae). J. Palynol. 25: 51–54.

Murthy, G.V.S. (1989) 1992. Palynotaxonomy of the genus *Butea* Roxburgh ex Willdenow (s.l.) (Fabaceae). Bull. Bot. Surv. India 31: 83–88.

Okolo, M.O. and Gill, L.S. 1986. Pollen morphological studies of the subfamily Papilionoideae (Leguminosae) from Southern Nigeria. J. Palynol. 22: 153–190.

Polhill, R.M. 1981. Papilionoideae. In: R.M. Polhill and P.H. Raven (eds.), Advances in Legume Systematics, part 1: 191–208. Royal Botanic Garden, Kew.

Punt, W., Blackmore, S., Nilsson, S. and Le Thomas, A. 1994. Glossary of pollen and spore terminology. LPP contribution series no. 1, LPP Foundation, Utrecht. 71 pp.

Ridder-Numan, J.W.A. 1992. *Spatholobus* (Leguminosae–Papilionoideae): a new species and some notes. Blumea 37: 63–71.

Ridder-Numan, J.W.A. and Kornet, D.J. 1994. A revision of the genus *Kunstleria* (Leguminosae–Papilionoideae). Blumea 38: 465–485.

Ridder-Numan, J.W.A. and Wiriadinata, H. 1985. A revision of the genus *Spatholobus* (Leguminosae–Papilionoideae). Reinwardtia 10: 139–205.

Sanjappa, M. 1987. Revision of the genera *Butea* Roxburgh ex Willdenow and *Meizotropis* Voigt (Fabaceae). Bull. Bot. Surv. India 29: 199–225.

Shome, U., Khanna, R.K. and Sharma, H.P. 1980. Pharmacognostic studies on the flower of *Butea monosperma* (Lamarck) Kuntze. New Botanist 7: 111–125.

Swofford, D.L. 1991. PAUP: Phylogenetic Analysis Using Parsimony, Version 3.0s. Computer-program formerly distributed by the Illinois Natural History Survey, Champaign, Illinois. 178 pp.

Tewari, R.B. and Nair, P.K.K. 1978. Apertural forms and their evolutionary trends in the pollen grains of the Indian Papilionaceae. Ind. J. Bot. 1: 133–138.

Tewari, R.B. and Nair, P.K.K. 1979. Pollen morphology of some Indian Papilionaceae. J. Palynol. 15: 49–73.

Van der Ham, R.W.J.M. 1990. Nephelieae pollen (Sapindaceae): form, function and evolution. Leiden Bot. Series 13, Leiden. 255 pp.

Vishnu-Mittre and Sharma, B.D.1962. Studies of Indian pollen grains 1. Leguminosae. Pollen et Spores 4: 5–45.

Legends of Plates I–V:

Plate I

Pollen of *Butea* (a–d), *Meizotropis* (e–h), and *Kunstleria* (i–o). SEM and TEM; bar = 10 μm (a, e, i, j, m) or 1 μm (b–d, f–h, k, l, n, o). e = endexine, f = foot layer, i = infratectum, t = tectum.

a. *Butea monosperma* (*Dickason 693960*). Oblique polar view. Note the obtuse colpus ends and the microreticulate tectum.

b. *B. monosperma* (*Kostermans 24393*). High magnification of the microreticulate tectum.

c. *B. superba* (*Gamble s.n.*). SEM fracture, showing exine stratification.

d. *B. monosperma* (*Kostermans 28597*). TEM section. Note the massive endexine, the irregular foot layer and the columellae.

e. *Meizotropis buteiformis* (*Ludlow et al. 20928*). Oblique equatorial view. Note the different ornamentation at the apocolpium.

f. *M. pellita* (*Stainton et al. 45*). High magnification of the verrucate ornamentation.

g. *M. pellita* (*Polunin et al. 1980*). SEM fracture near aperture, showing exine stratification.

h. *M. buteiformis* (*Koelz 28394*). TEM section. Note the massive endexine and the very thin foot layer.

i. *Kunstleria forbesii* (*Elmer 21322*). Polar view, showing acute colpus ends.

j. *K. ridleyi* (*Ahmed 1226*). Equatorial view, showing colpate ectoaperture.

k. *K. forbesii* (*Forbes 3241*). TEM section. Note the irregular columellae.

l. *K. ridleyi* (Hort. Bog. XVII.F.13). TEM section. Note the regular, well-spaced columellae.

m, n. *K. philippinensis* (*Wenzel 818*). – m. Oblique view. – n. High magnification of the rugulate tectum.

o. *K. sarawakensis* (*S 18036*). SEM fracture. Note the endexine thickening towards the ectoaperture.

Plate II

Pollen of *Spatholobus*, Type I (a–c) and Type II (d–m). SEM and TEM; bar = 10 µm (a, d, e, g, k) or 1 µm (b, c, f, h–j, l, m). e = endexine, f = foot layer, i = infratectum, t = tectum.

a, b.　*S. parviflorus* (*Koelz 19167*). – a. Oblique polar view. Note the long colpi with acute ends. – b. High magnification of the perforate, slightly fossulate tectum.

c.　*S. parviflorus* (*Garret 1353*). TEM section. Note the massive endexine, the very thin foot layer and the irregular, granular columellae.

d–f.　*S. apoensis* (*Elmer 13300*). – d. Polar view, showing normal ornamentation, long colpi. – e. Oblique view, showing deviating coarsely perforate-fossulate to verrucate ornamentation. Note the obtuse colpus ends. – f. TEM section, showing irregular columellae and coarse, deviating tectum structure.

g.　*S. maingayi* (*Santos 4064*). Polar view, showing acute colpus ends and microreticulate tectum.

h.　*S. maingayi* (*PNH 2911*). High magnification of the microreticulate tectum, showing one free sexine element per lumen.

i.　*S. littoralis* (*Winckel 325*). TEM section. Note the coarsely granular infratectum.

j.　*S. maingayi* (*Kunstler 10428*). TEM section of two adjacent exine parts (one upside down), showing irregular columellae and normal tectum structure.

k, l.　*S. latistipulus* (*Elmer 21439*). – k. Oblique view. – l. High magnification of the coarsely fossulate to microreticulate tectum.

m.　*S. maingayi* (*SAN 76946*). SEM fracture, showing exine stratification. Note the probable foot layer / endexine border.

Plate III

Pollen of *Spatholobus*, Subtype IIIa (a–i, l–n) and Subtype IIIb (j, k; see also Plates IV and V, a–c). SEM and TEM; bar = 10 µm (a, e, f, i, j) or 1 µm (b–d, g, h, k–n). e = endexine, f = foot layer, i = infratectum, t = tectum.

a, b.　*S. ferrugineus* (*Elsener H88*). – a. Oblique view. – b. High magnification of the psilate-perforate tectum.

c.　*S. ferrugineus* (*van Niel 3868*). SEM fracture. Note the regular, well-spaced columellae.

d.　*S. ferrugineus* (*van Balgooy 3977*). TEM section. Note the channels in the foot layer.

e.　*S. gyrocarpus* (*KL 2758*). Polar view.

f, h.　*S. macropterus* (*Forbes 3243*). – f. Polar view. – h. TEM section. Note the slightly discontinuous foot layer.

g.　*S. macropterus* (*FRI 10529*). SEM fracture. Note the slightly less regular columellae (compare with Plate III, c).

i, l.　*S. hirsutus* (*Nieuwenhuis 1079*). – i. Equatorial view. – l. High magnification of the micro-rugulate tectum.

j.　*S. multiflorus* (*Maidin 7331*). Equatorial view, showing psilate-perforate ornamentation.

k.　*S. multiflorus* (*SAN 110486*). TEM section.

m.　*S. hirsutus* (*SAN 37132*). SEM fracture, showing the more or less regular columellae.

n.　*S. hirsutus* (*Mogea 4085*). TEM section. Note the channels in the foot layer.

Plate IV

Pollen of *Spatholobus*, Subtype IIIb (see also Plates III, j, k and V, a–c). SEM and TEM; bar =
10 µm (a, b, e, f, i, j, n) or 1 µm (c, d, g, h, k, l, o). e = endexine, f = foot layer, i = infratectum,
t = tectum.

a, b.   *S. acuminatus* (*Poilane 14573*). – a. Polar view. – b. Equatorial view.

c.   *S. acuminatus* (*Kerr 18291*). SEM fracture, showing irregular columellae.

d.   *S. acuminatus* (*Maxwell 75-137*). TEM section, showing exine stratification.

e, f.   *S. albus* (*S 32363*). – e. Oblique polar view. Note connection between two colpi. –
f. Equatorial view.

g.   *S. harmandii* (*How 71016*). SEM fracture, showing infratectum consisting of coarse
granules.

h.   *S. albus* (*S 32363*). TEM section, showing exine stratification.

i, j.   *S. oblongifolius* (*SAN 43060*). – i. Polar view. Note the long colpi and the acute colpus
ends. – j. Equatorial view.

k.   *S. oblongifolius* (*SAN 65944*). SEM fracture, showing exine stratification.

l.   *S. oblongifolius* (*SAN 100088*). TEM section. Note the granule-like columellae.

m.   *S. purpureus* (*Stocks s.n.*). SEM section. Note the inside of the endoaperture.

n.   *S. viridis* (*SAN 97423*). Polar view.

o.   *S. viridis* (*SAN 66270*). TEM section. Note the irregular short columellae and the discontinuous foot layer.

Plate V

Pollen of *Spatholobus*, Subtype IIIb (a–c) and Subtype IIIc (d–m). SEM and TEM; bar = 10 µm
(a, b, d, e, h, k) or 1 µm (c, f, g, j, l, m). e = endexine, f = foot layer, i = infratectum, t = tectum.

a, c.   *S. persicinus* (*Haviland 3276*). – a. Polar view of syncolporate pollen grain. – c. High
magnification of the fossulate tectum.

b.   *S. persicinus* (*Kostermans 7108*). Oblique equatorial view.

d.   *S. auritus* (*Haniff 3703*). Oblique polar view. Note the obtuse colpus ends.

e.   *S. suberectus* (*Bor 5067*). Equatorial view, showing acute colpus ends and granules along
the mesocolpia.

f.   *S. suberectus* (*Spire 453*). SEM fracture, showing dense infratectal structure.

g.   *S. suberectus* (*Henry 11977A*). TEM section, showing exine stratification.

h, j.   *S. latibractea* (*S 45351*). – h. Oblique polar view. Note the almost microreticulate tectum. – j. TEM section, showing irregular granule-like columellae.

i.   *S. latibractea* (*SAN 97466*). Equatorial view, showing the psilate tectum.

k.   *S. pottingeri* (*BKF 37401*). Oblique view. Note the acute colpus ends.

l.   *S. pulcher* (*Henry 12780A*). High magnification of the psilate tectum.

m.   *S. merguensis* (*Parkinson 1690*). SEM section, showing exine stratification.

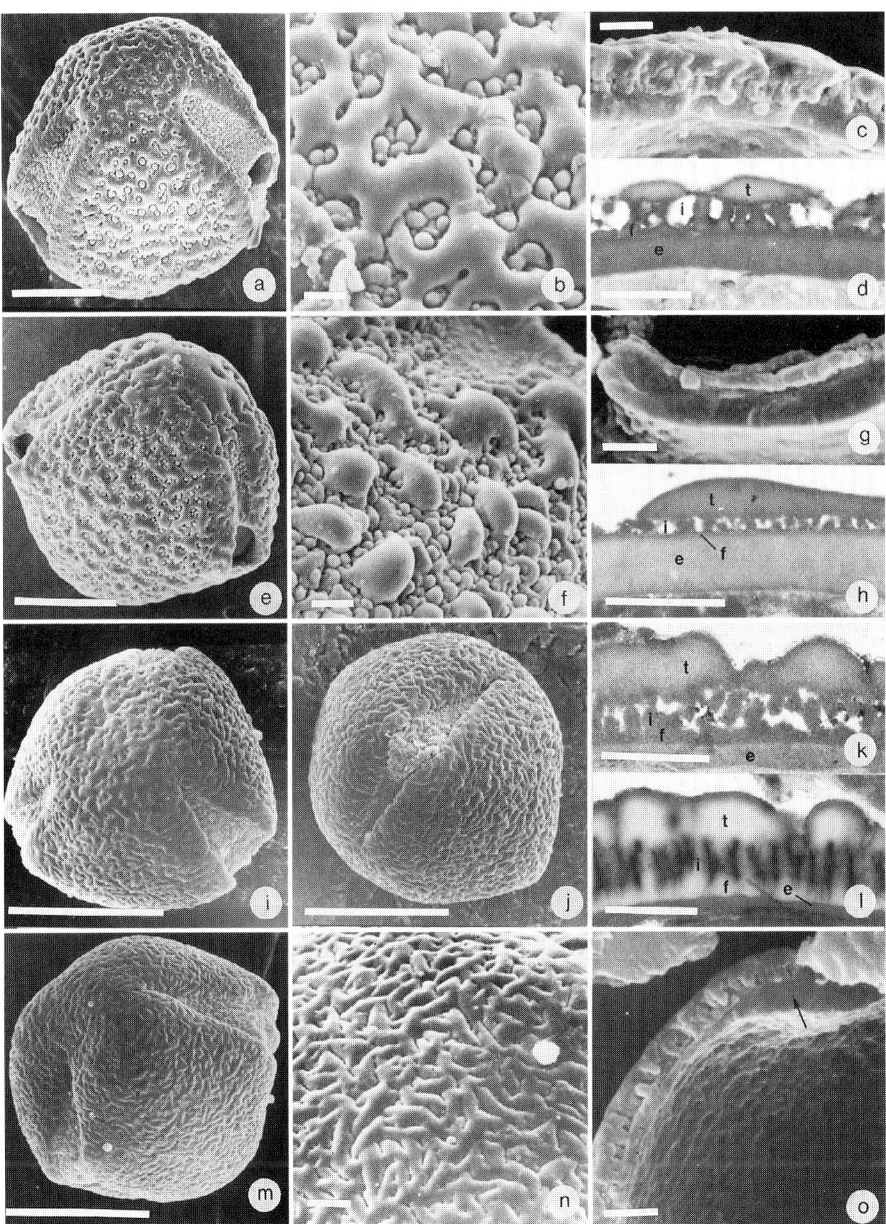

Plate I — For the legend, see page 166.

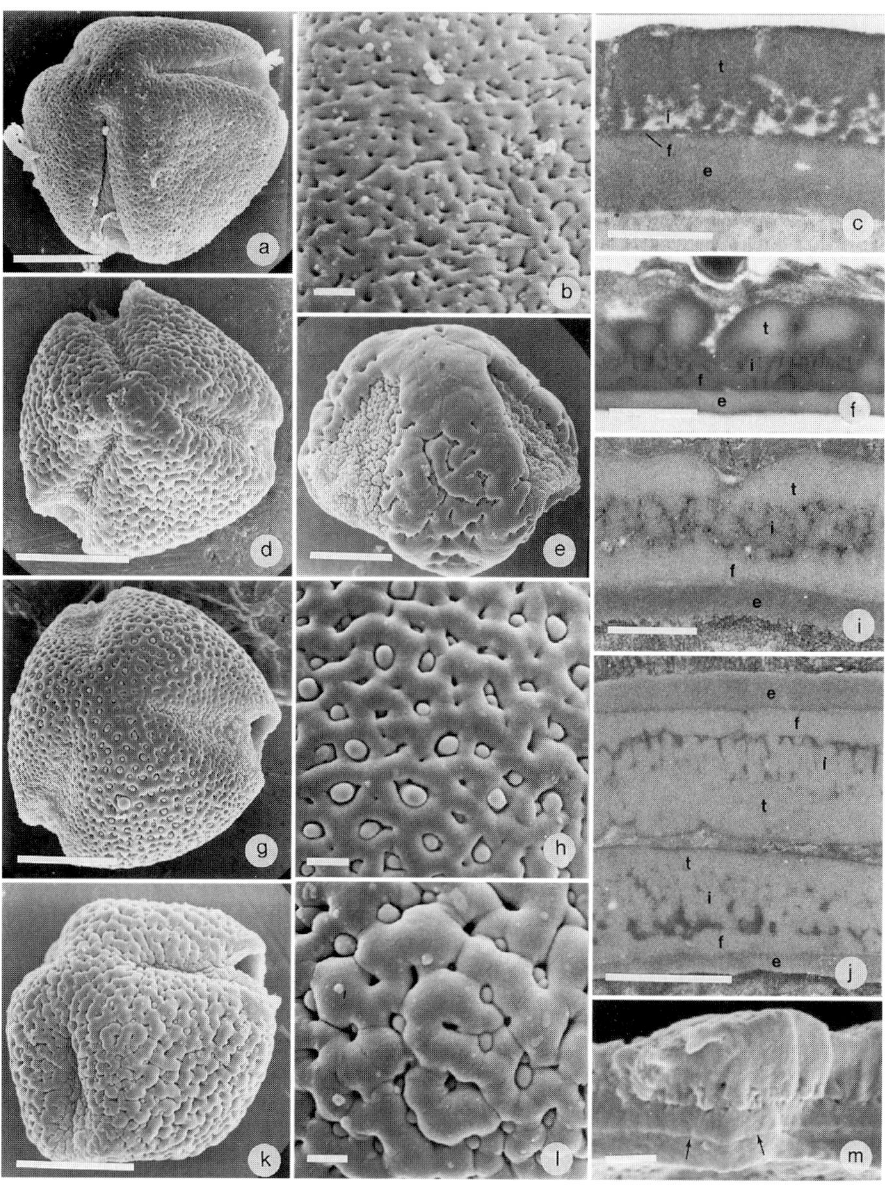

Plate II — For the legend, see page 167.

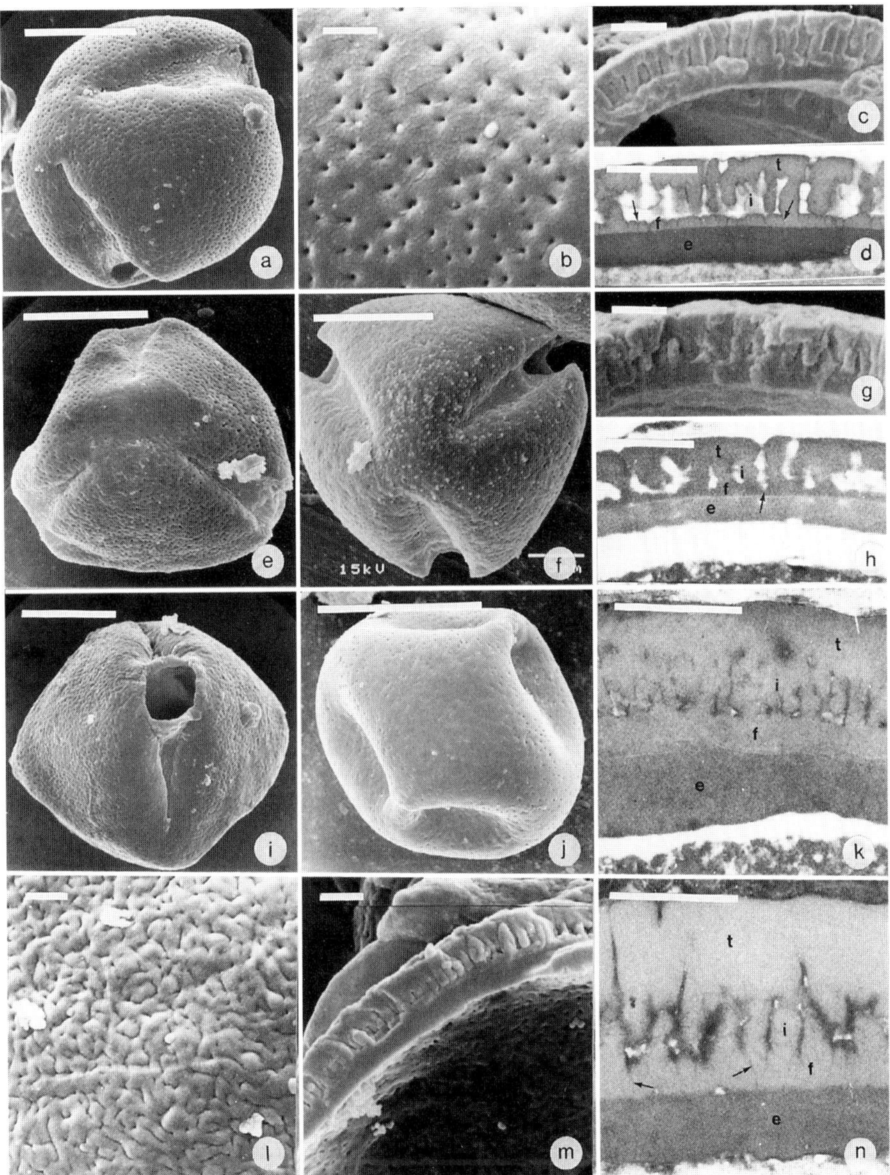

Plate III — For the legend, see page 167.

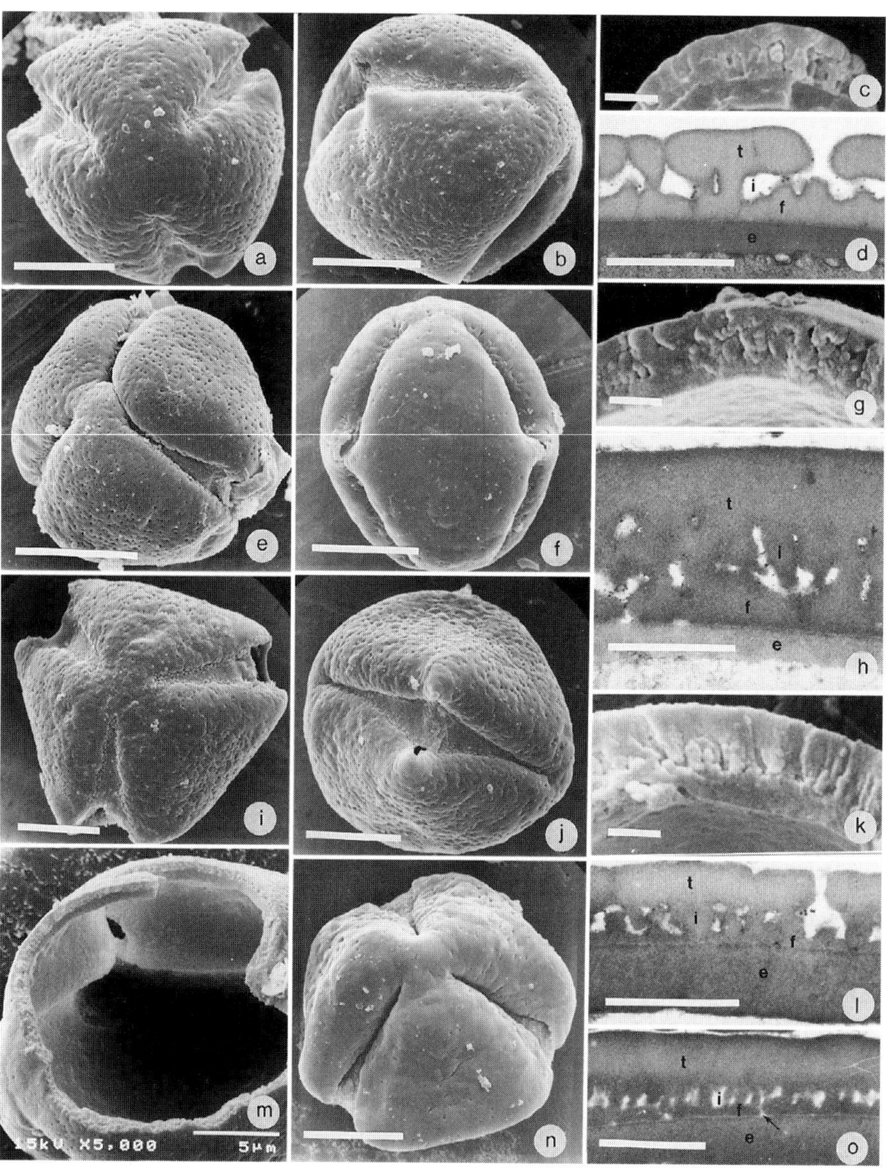

Plate IV — For the legend, see page 168.

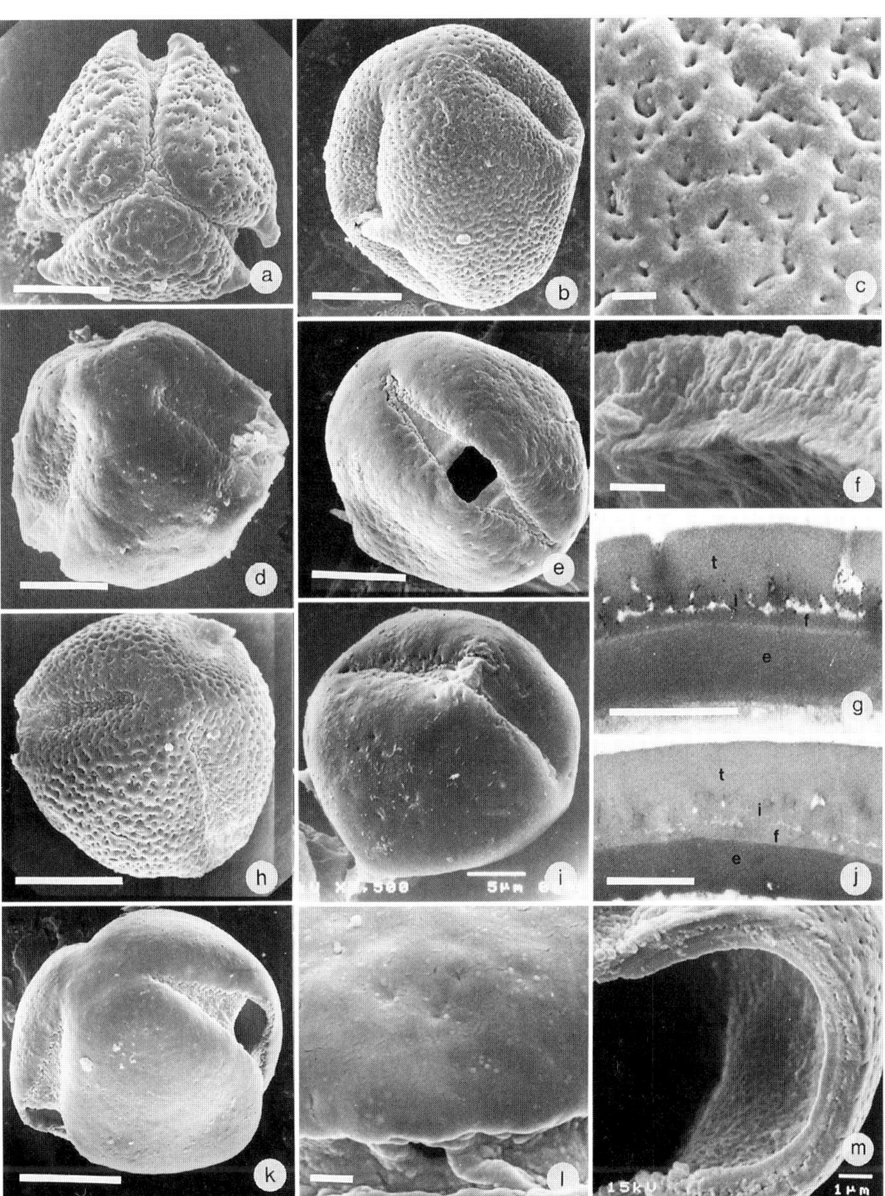

Plate V — For the legend, see page 168.

SAMENVATTING

Het doel van dit proefschrift was het onderzoeken van de samenhang tussen enerzijds de verwantschap binnen een groep planten en de verspreiding van deze planten en anderzijds de geologische historie van het gebied waar deze planten voorkomen. Kortom, kun je iets zeggen over de historie van de gebieden of over de verspreidingsgeschiedenis van een groep planten als je, behalve de verwantschappen van de planten onderling, ook de verspreiding van de planten kent. De hoofdgedachte achter het historisch-biogeografische onderzoek is dat planten die nauwer aan elkaar verwant zijn ook dichter bij elkaar voorkomen. Eén van de aanleidingen tot het ontstaan van twee soorten uit één voorouder is het opsplitsen van gebieden ('areas'). Hierdoor ontstaan uit de oorspronkelijke voorouder twee van elkaar geïsoleerde groepen. Deze twee groepen kunnen in de loop der tijd hun eigen soortsvormingsproces ondergaan. Gebieden kunnen door verschillende oorzaken worden opgesplitst, zoals gebergtevorming, zeespiegelstijging (waardoor lager gelegen gebieden onderlopen), klimaatsverandering (IJstijden) en het uit elkaar drijven van de continenten.

De groep planten die werd onderzocht is het geslacht *Spatholobus*, een groep van 29 lianen die behoren tot de familie van de Leguminosae, onderfamilie Papilionoideae (vlinderbloemigen), tribus Phaseoleae. Twee nauw verwante geslachten, *Butea* en *Meizotropis*, beide met twee soorten, werden in het onderzoek meegenomen. Deze drie geslachten hebben altijd een in drie blaadjes gedeeld blad, en peulen met één zaad en een vleugel.

Om de verwantschappen binnen deze groep planten te weten te komen werd er een fylogenetische analyse uitgevoerd. Bij deze methode, die het eerst uitgebreid is beschreven door Hennig, gaat men er van uit dat planten (of dieren) in de loop der tijd zijn veranderd, waarbij zij uit de gezamenlijke eigenschappen van de voorouders (plesiomorfe kenmerken) nieuwe varianten hebben ontwikkeld (apomorfe of afgeleide kenmerken). Het gemeenschappelijk bezit van dergelijke afgeleide kenmerken is te gebruiken voor het vaststellen van verwantschappen. In een soort stamboom, een cladogram, worden de soorten aan de uiteinden van de takken gezet. De basale tak stelt de voorouder van de gehele groep voor, met een aantal 'primitieve' kenmerken. Uit deze voorouder ontstaan nakomelingen, die weliswaar nog een groot aantal plesiomorfe kenmerken gemeenschappelijk hebben met de voorouder, maar die ook hun eigen apomorfe kenmerkstoestanden vertonen. Uit deze nakomelingen ontstaat weer nieuw nageslacht, en zo blijft het cladogram zich vertakken tot op het soortsniveau van de nu levende planten.

Omdat aan een groep planten niet zonder meer te zien is welke kenmerkstoestanden apomorf of plesiomorf zijn, wordt een zogenaamde 'outgroup' toegevoegd. Dit is een (groep) soort(en), die nauw genoeg verwant is aan de te onderzoeken groep om daarmee zinvol vergeleken te kunnen worden, maar waarvan ook bekend is dat ze over meer 'primitieve' kenmerkstoestanden beschikt. Voor dit onderzoek werd het geslacht *Kunstleria* als outgroup gekozen. Deze lianen horen tot het tribus Millettieae, dat wordt beschouwd als een minder afgeleide groep dan het tribus Phaseoleae. Het geslacht *Spatholobus* zit waarschijnlijk op het grensvlak van deze beide stammen.

Van alle soorten binnen de vier geslachten werden zoveel mogelijk eigenschappen bekeken, en hun verschillende kenmerkstoestanden werden opgenomen in een tabel (datamatrix). De kenmerken die in dit onderzoek gebruikt zijn staan beschreven in Hoofdstuk 2. Er zijn kenmerken gebruikt die aan de buitenkant van de plant (of hooguit binnenin de bloem of vrucht) te zien zijn, maar ook kenmerken van de bladanatomie en de pollenmorfologie. Aan deze laatste categorie is een apart artikel gewijd, dat hier is opgenomen als Hoofdstuk 5.

De datamatrix werd met behulp van speciale computerprogramma's geanalyseerd. Het resultaat, drie cladogrammen die de meest waarschijnlijke verwantschapsrelaties tussen de soorten van *Spatholobus* c.s aangeven, staat in Figuur 2.25. Uit deze drie werd er uiteindelijk één geselecteerd om mee verder te werken.

Het gebied waar de onderzochte geslachten voorkomen, is Zuidoost-Azië. De geslachten *Butea* en *Meizotropis* komen alleen op het vasteland voor, *Spatholobus* komt ook voor in de Maleise Archipel (Fig. 3.1). Dit deel van de wereld is in geologisch opzicht een heel complex gebied, omdat het op het grensvlak ligt van een aantal platen in de aardkorst (Indo-Australische, Euraziatische, Pacifische en Filippijnse Zee plaat). Deze platen zijn voortdurend in beweging, waarbij ze van elkaar of naar elkaar toe kunnen drijven. In het eerste geval zal op het breukvlak vloeibaar gesteente (lava) omhoog komen en stollen (oceaanruggen), in het andere geval botsen de platen op elkaar, of schuiven ze over elkaar heen (subductie).

In de loop der tijden, vanaf het Paleozoïcum, zijn er vanuit het zuiden brokstukken van de rand van het grote continent Gondwanaland afgebroken en naar het noorden gedreven. Deze brokstukken botsten uiteindelijk tegen de Euraziatische plaat aan, en vormden met elkaar onder andere Zuidoost-Azië. Eén van de bekendste brokstukken is India, dat omstreeks 50 miljoen jaar geleden tegen de Euraziatische plaat botste, waarna de Himalaya begon te ontstaan. Er zijn in dit gebied niet alleen brokstukken met een continentale oorsprong te vinden, er zijn ook gebieden die ontstaan zijn door vulkanisme op de rand van een continentale plaat (bijvoorbeeld een groot deel van Java en de eilanden ten oosten daarvan).

De aanwezigheid van een bepaald gebied geeft nog geen garantie dat er ook plantengroei mogelijk was: gedurende bepaalde perioden (85–55 miljoen jaar geleden) was de zeespiegel zelfs 350 meter boven het huidige niveau. Echter, de zeespiegel is ook veel lager geweest dan nu, bijvoorbeeld tijdens de ijstijden in het Pleistoceen. Tijdens dit soort perioden lag het Sundaplateau, tussen Borneo en het Maleise schiereiland, droog. Naast direct geologische zijn uiteraard ook klimatologische factoren van belang voor het voorkomen van organismen in een gebied. In Hoofdstuk 3 wordt een samenvatting gegeven van de geologie van dit gebied.

In Hoofdstuk 4 worden de resultaten van de fylogenetische analyse (cladogram b uit Fig. 2.25) gebruikt voor een cladistische biogeografische analyse. Voor een biogeografische analyse worden min of meer dezelfde computerprogramma's gebruikt als voor de fylogenetische. De gegevens voor de datamatrix zijn in dit geval de verspreiding van de verschillende soorten en hun voorouders. Om de verspreiding van de voorouders te bepalen werd uitgegaan van het cladogram. Er werd, vanuit methodologische overwegingen, verondersteld dat de voorouder dezelfde verspreiding heeft als de nakomelingen tezamen. In feite gaat de methode er dus van uit dat de voorouder van de groep in het gehele gebied voorkwam. Dit gebied zou dan in stukjes zijn opgesplitst, waarna in elk stukje na elke splitsing een nieuwe soort ontstond. Een voor dit soort onderzoek ideale groep zou zich dus niet hebben verplaatst nadat de voorouder zijn positie had ingenomen, en bovendien op elke splitsing van een gebied gereageerd moeten hebben door middel van soortsvorming. Uiteraard is dit slechts theorie, en is het goed mogelijk dat de planten zich niet zo netjes aan de grenzen van het voorouderareaal hebben gehouden. In gunstige omstandigheden zullen ze hun verspreidingsgebied hebben uitgebreid en in ongunstige hebben ingekrompen. Dit laatste, de voor de groep specifieke reacties op geologische en klimatologische omstandigheden, kunnen echter wel ontdekt worden doordat ze een afwijkend beeld geven van de verspreiding van de groep binnen een meer algemeen patroon dat door meerdere groepen tezamen wordt gegeven.

Als blijkt dat meerdere groepen planten (en dieren) een bepaald verspreidingspatroon hebben, kan er sprake zijn van een algemeen patroon dat bijvoorbeeld door een speciale geologische gebeurtenis is veroorzaakt.

In de historische biogeografie is men in feite op zoek naar deze geologische gebeurtenissen, de 'splitsingen' van gebieden (zoals het ontstaan van bergketens, isolatie van gebieden door stijgende zeespiegel) die gevolgd worden door soortsvorming. Om te zorgen dat de uitkomsten zo algemeen mogelijk zijn, werden, behalve de gegevens van *Spatholobus* en verwanten, ook de gegevens gebruikt van door andere onderzoekers gepubliceerde resultaten van fylogenetische analyses (*Fordia*, *Genianthus*, *Xanthophytum*). Deze gegevens werden gezamenlijk geanalyseerd.

Het uiteindelijke resultaat van de biogeografische analyses (Fig. 4.8) is een hypothese over de historische relaties tussen de gebieden (een algemeen areacladogram). In het areacladogram is te zien dat er een splitsing is tussen de gebieden op het vasteland van Zuidoost-Azië en de gebieden ten zuiden van de Isthmus van Kra. Verder is er een splitsing tussen de gebieden op Borneo en de overige gebieden in de Maleise Archipel. Blijkbaar is er een historische oorzaak voor de scheiding tussen deze groepen. Het zou bijvoorbeeld mogelijk kunnen zijn dat er een splitsing heeft plaatsgevonden tussen het vasteland en het gedeelte ten zuiden van de Isthmus van Kra. Maar het is waarschijnlijker dat deze splitsing te maken heeft met het droogvallen van stukken land, waardoor soorten die voor die tijd uitsluitend op het vasteland voorkwamen, konden migreren naar meer zuidelijk gelegen delen, die vóór die tijd niet te bereiken waren (ook de Isthmus van Kra bevond zich herhaaldelijk onder water). Deze uitleg wordt ondersteund door onder andere de verspreiding van fossielen van de Papilionoideae, die pas vrij laat in de tijd in Zuidoost-Azië werden gevonden, en in noordelijke gebieden al eerder. Ook is te zien in het areacladogram dat de gebieden op Borneo in een bepaalde volgorde aftakken: de oudste gebieden takken vrij snel af, de jongere gebieden pas later.

In het tweede deel van Hoofdstuk 4 worden de resultaten van de biogeografische analyse vergeleken met de geologische gegevens uit Hoofdstuk 3, en wordt de mogelijke verspreidingsgeschiedenis van *Spatholobus* gereconstrueerd. Het is het meest waarschijnlijk dat deze groep ontstond op het vasteland van Zuidoost-Azië, en gedurende periodes van laag water en een droger klimaat de Maleise Archipel bereikte en het Sundaplateau overstak naar Borneo. Omdat verschillende zijtakken van het cladogram telkens weer opnieuw een soortgelijk patroon laten zien, is dit waarschijnlijk in de loop van de tijd meer dan eens gebeurd, waarbij er telkens nieuwe elementen aan de al bestaande flora werden toegevoegd. Het is echter heel speculatief om een tijdschaal aan de verschillende oversteken toe te voegen, omdat er vanaf het Eoceen regelmatig langere perioden met hoog en laag water zijn geweest. De eerste oversteek van *Spatholobus* naar de toenmalige Maleise Archipel zou dus op zijn vroegst gedateerd kunnen worden in het Eoceen.

CURRICULUM VITAE

Jeannette Wilma Abéle Numan werd geboren op 22 januari 1957 te Emmen. In 1975 behaalde zij het Gymnasium-b diploma aan de Christelijke Scholengemeenschap te Voorburg, waarna zij biologie ging studeren aan de Rijksuniversiteit te Leiden. Op 26 september 1978 legde zij het kandidaatsexamen B1w af (biologie met als bijvak mathematische statistiek). Aansluitend werd de doctoraalfase ingevuld met onder andere de volgende onderdelen:

- Van 1978 tot 1980 op het Rijksherbarium: Een revisie van het geslacht *Spatholobus* (Leguminosae–Papilionoideae), onder leiding van dr. R. Geesink.
- Van 1980 tot 1983 op het Laboratorium voor Experimentele Plantensystematiek: De relatie tussen het galmuggen geslacht *Dasyneura* en haar gastheren onder begeleiding van dr. H. C. Roskam. Dit onderzoek werd gedurende anderhalf jaar onderbroken voor, respectievelijk, de Experimentele Lerarenopleiding, een aansluitende baan voor een half jaar als lerares biologie op Christelijke Scholengemeenschap Overvoorde te Den Haag en een half jaar waarin zij thuis bleef vanwege zwangerschap en de geboorte van een zoon.
- Van 1981 tot 1982 de Experimentele Lerarenopleiding, een universitaire cursus met zes maanden full-time stage onder leiding van E. Herbschleb op het Westland College te Naaldwijk en het Christelijk Gymnasium Sorghvliet te Den Haag.

In 1981 was zij studentassistent bij de cursussen Overzicht Plantenrijk voor Farmaceuten en Planten Systematiek voor Biochemici. In 1983 assisteerde zij tijdens het 1e-jaars practicum Overzicht Plantenrijk.

Het doctoraalexamen werd op 30 augustus 1983 afgelegd, waarna zij aansluitend in dienst kwam als bevoegd lerares biologie bij de Christelijke Scholengemeenschap Overvoorde te Den Haag. Hier bleef zij met veel plezier lesgeven tot augustus 1989. Tijdens deze periode werden een tweede en derde zoon geboren.

Vanaf 1 september 1989 tot 16 februari 1995 volgde een part-time aanstelling als onderzoeker in opleiding (oio) bij NWO, met als standplaats het Rijksherbarium te Leiden. Het onderzoek op dit instituut, dat uiteindelijk leidde tot het proefschrift, werd begeleid door dr. R. Geesink. Na zijn onverwacht overlijden in 1992 werd de begeleiding overgenomen door dr. M.C. Roos.

Op dit moment is zij verbonden als gastmedewerker aan het onderzoekinstituut Rijksherbarium / Hortus Botanicus.

Publicaties:

Ridder-Numan, J.W. A. & H. Wiriadinata. 1985. A revision of the genus *Spatholobus* (Leguminosae–Papilionoideae). Reinwardtia 10 (2): 139–205.

Ridder-Numan, J.W. A. 1992. *Spatholobus* (Leguminosae–Papilionoideae): a new species and some taxonomical notes. Blumea 37: 63–71.

Ridder-Numan, J.W. A. & D.J. Kornet. 1994. A revision of the genus *Kunstleria* (Leguminosae–Papilionoideae). Blumea 38: 465–485.

Ridder-Numan, J.W. A. 1995. Phylogeny and biogeography of *Spatholobus, Butea, Meizotropis* and *Kunstleria* (Legum.-Pap.). In: M. Crisp & J.J. Doyle (eds.), Advances in Legume Systematics, Part 7: Phylogeny: 133–139. Royal Botanic Gardens, Kew.

Ridder-Numan, J.W. A. 1996. Phylogeny and biogeography of *Spatholobus* and allies (Legum.-Pap.): just another case history. Austral. J. Syst. Bot. 9 (2): 127–132.

Ridder-Numan, J.W. A. & R.W. J.M. van der Ham. Pollen morphology of *Butea, Kunstleria, Meizotropis*, and *Spatholobus* (Leguminosae–Papilionoideae), with notes on their position in the tribes Millettieae and Phaseoleae. Review of Palaeobotany and Palynology (geaccepteerd). Dit proefschrift Chapter 5.

Presentaties op internationale symposia:

Third International Legume Conference, Kew, 12–17 juli 1992. Poster: Phylogeny and biogeography of *Spatholobus, Butea* and *Kunstleria* (Leguminosae–Papilionoideae).

XIth Meeting of the Willy Hennig Society, Parijs, 24–28 augustus 1992. Poster: Phylogeny and biogeography of *Spatholobus, Butea* and *Kunstleria* (Leguminosae–Papilionoideae).

Second International Flora Malesiana Symposium, Yogyakarta, 6–12 september 1992. Lezing: Phylogeny and biogeography of *Spatholobus, Butea* and *Kunstleria* (Leguminosae–Papilionoideae).

Origin and Evolution of the Flora of the Monsoon Tropics (Symposium Australian Syst. Bot. Society), Kuranda, 4–6 juli 1994. Lezing: Phylogeny and biogeography of *Spatholobus* and allies (Leguminosae–Papilionoideae): just another case history.

XIIIth Meeting of the Willy Hennig Society, Kopenhagen, 23–26 augustus 1994. Lezing: Phylogeny and biogeography of *Spatholobus* and allies (Leguminosae–Papilionoideae): just another case history.

Third International Flora Malesiana Symposium, Kew, 9–14 juli 1995. Lezing: The continuing story of *Spatholobus* (Leguminosae–Papilionoideae) and allies in Southeast Asia.

Biogeography and Geological Evolution of Southeast Asia, London, 6–7 maart 1996. Poster: Historical biogeography of *Spatholobus* and allies (Leguminosae–Papilionoideae) in Southeast Asia.

NAWOORD

Het onderzoek dat resulteerde in dit proefschrift duurde van september 1989 tot februari 1995. Zonder dr. Rob Geesink, die ook mijn afstudeeronderzoek begeleidde van 1978 tot 1980, zou dit onderzoek er nooit geweest zijn. Hij wist mij uit de school te lokken en over te halen om, na meer dan vijf jaar, het wetenschappelijk onderzoek weer op te vatten. Hij bracht met veel enthousiasme iedereen die dat wilde de beginselen van de fylogenie bij. Helaas overleed hij vroegtijdig in september 1992.

Het onderzoekinstituut Rijksherbarium/Hortus Botanicus is een instituut met goede onderzoeksfaciliteiten en een plezierige werksfeer. Ook was het een voorrecht van de uitgebreide collecties gebruik te kunnen maken. Het is voor een groot deel ook aan mijn collega's te danken dat dit proefschrift voor u ligt en ik ben hen, vooral de onderzoekgroep Tropische Fanerogamen, erkentelijk voor hun gewillig oor.

Marieke Hardenberg maakte met veel zorg de preparaten die nodig waren voor het pollen- en bladanatomisch onderzoek, en we verrichtten samen een groot deel van het scanwerk. Ook Bertie Joan van Heuven stond altijd klaar om nog wat preparaten te maken of foto's af te drukken. Wim Star maakte de TEM-preparaten en foto's, die onontbeerlijk waren voor het pollenonderzoek. De tekenaars Jan van Os en Joop Wessendorp hebben een aantal mooie tekeningen gemaakt voor een deel van dit proefschrift.

Emmy van Nieuwkoop heeft met enthousiasme en energie een deel van haar tijd in dit proefschrift gestoken. Met grote vakkundigheid verzorgde zij de lay-out. Het was ook een hele ervaring met haar samen te werken aan de elektronische verwerking van een groot aantal van de figuren. Ik ben Jos Rietstap (TNO-IAG, Delft) zeer erkentelijk voor zijn bijdrage in het scannen van illustraties, en de vele adviezen met betrekking tot het maken van de figuren in de computer.

Tiny Verberne en Ginnie Heerdink, de beide trouwe oppassen van Pieter, Mark en Bouke, maakten het mogelijk met een gerust hart het onderzoek op me te nemen en af te ronden. Zonder hun inzet was dat niet mogelijk geweest.

Mijn ouders, die tevens als reserve-oppas fungeerden, ben ik dankbaar voor hun steun op velerlei gebied.

Jan, die mij vaak ongemerkt met raad en daad ter zijde stond, wil ik bedanken voor zijn geduld en steun. Mede door hem is het mogelijk geworden dit proefschrift op zo elektronisch mogelijke wijze in elkaar te zetten.

Het onderzoek werd gefinancierd door de Stichting Levenswetenschappen (SLW) van de Nederlandse Organisatie voor Wetenschappelijk Onderzoek (NWO) onder projectnummer 805-40-041. Deze stichting maakte ook de meeste van mijn congresbezoeken mogelijk.

De Alberta Mennega Stichting betaalde de kosten voor een congres- en werkbezoek aan het Herbarium in Kew, Engeland.